Big Data Analytics for Satellite Image Processing and Remote Sensing

P. Swarnalatha
VIT University, India

Prabu Sevugan
VIT University, India

A volume in the Advances in Computer and Electrical Engineering (ACEE) Book Series

IGI Global
DISSEMINATOR OF KNOWLEDGE

Published in the United States of America by
IGI Global
Engineering Science Reference (an imprint of IGI Global)
701 E. Chocolate Avenue
Hershey PA, USA 17033
Tel: 717-533-8845
Fax: 717-533-8661
E-mail: cust@igi-global.com
Web site: http://www.igi-global.com

Library of Congress Cataloging-in-Publication Data

Names: Swarnalatha, P. (Purushotham), 1977- editor. | Sevugan, Prabu, 1981- editor.
Title: Big data analytics for satellite image processing and remote sensing / P. Swarnalatha and Prabu Sevugan, editors.
Description: Hershey, PA : Engineering Science Reference, [2018] | Includes bibliographical references.
Identifiers: LCCN 2017019875| ISBN 9781522536437 (hardcover) | ISBN 9781522536444 (ebook)
Subjects: LCSH: Geospatial data. | Big data. | Artificial satellites in remote sensing. | Image processing--Digital techniques. | Artificial satellites--Optical observations--Data processing.
Classification: LCC GA102.4.E4 B54 2018 | DDC 621.36/7028557--dc23 LC record available at https://lccn.loc.gov/2017019875

This book is published in the IGI Global book series Advances in Computer and Electrical Engineering (ACEE) (ISSN: 2327-039X; eISSN: 2327-0403)

British Cataloguing in Publication Data
A Cataloguing in Publication record for this book is available from the British Library.

All work contributed to this book is new, previously-unpublished material.
The views expressed in this book are those of the authors, but not necessarily of the publisher.

For electronic access to this publication, please contact: eresources@igi-global.com.

Advances in Computer and Electrical Engineering (ACEE) Book Series

ISSN:2327-039X
EISSN:2327-0403

Editor-in-Chief: Srikanta Patnaik, SOA University, India

MISSION

The fields of computer engineering and electrical engineering encompass a broad range of interdisciplinary topics allowing for expansive research developments across multiple fields. Research in these areas continues to develop and become increasingly important as computer and electrical systems have become an integral part of everyday life.

The **Advances in Computer and Electrical Engineering (ACEE) Book Series** aims to publish research on diverse topics pertaining to computer engineering and electrical engineering. **ACEE** encourages scholarly discourse on the latest applications, tools, and methodologies being implemented in the field for the design and development of computer and electrical systems.

COVERAGE

- Analog Electronics
- Qualitative Methods
- Power Electronics
- Programming
- Digital Electronics
- Electrical Power Conversion
- Algorithms
- Computer Architecture
- Computer Hardware
- VLSI Design

IGI Global is currently accepting manuscripts for publication within this series. To submit a proposal for a volume in this series, please contact our Acquisition Editors at Acquisitions@igi-global.com or visit: http://www.igi-global.com/publish/.

The Advances in Computer and Electrical Engineering (ACEE) Book Series (ISSN 2327-039X) is published by IGI Global, 701 E. Chocolate Avenue, Hershey, PA 17033-1240, USA, www.igi-global.com. This series is composed of titles available for purchase individually; each title is edited to be contextually exclusive from any other title within the series. For pricing and ordering information please visit http://www.igi-global.com/book-series/advances-computer-electrical-engineering/73675. Postmaster: Send all address changes to above address.

Titles in this Series

For a list of additional titles in this series, please visit:
https://www.igi-global.com/book-series/advances-computer-electrical-engineering/73675

EHT Transmission Performance Evaluation Emerging Research and Opportunities
K. Srinivas (Transmission Corporation of Andhra Pradesh Limited, India) and R.V.S. Satyanarayana (Sri Venkateswara University College of Engineering, India)
Engineering Science Reference •©2018 • 160pp • H/C (ISBN: 9781522549413) • US $145.00

Fuzzy Logic Dynamics and Machine Prediction for Failure Analysis
Tawanda Mushiri (University of Johannesburg, South Africa) and Charles Mbowhwa (University of Johannesburg, South Africa)
Engineering Science Reference •©2018 • 301pp • H/C (ISBN: 9781522532446) • US $225.00

Creativity in Load-Balance Schemes for Multi/Many-Core Heterogeneous Graph Computing ...
Alberto Garcia-Robledo (Center for Research and Advanced Studies of the National Polytechnic Institute (Cinvestav-Tamaulipas), Mexico) Arturo Diaz-Perez (Center for Research and Advanced Studies of the National Polytechnic Institute (Cinvestav-Tamaulipas), Mexico) and Guillermo Morales-Luna (Center for Research and Advanced Studies of the National Polytechnic Institute (Cinvestav-IPN), Mexico)
Engineering Science Reference • ©2018 • 217pp • H/C (ISBN: 9781522537991) • US $155.0

Free and Open Source Software in Modern Data Science and Business Intelligence ...
K.G. Srinivasa (CBP Government Engineering College, India) Ganesh Chandra Deka (M. S. Ramaiah Institute of Technology, India) and Krishnaraj P.M. (M. S. Ramaiah Institute of Technology, India)
Engineering Science Reference •©2018 • 189pp • H/C (ISBN: 9781522537076) • US $190.00

Design Parameters of Electrical Network Grounding Systems
Osama El-Sayed Gouda (Cairo University, Egypt)
Engineering Science Reference •©2018 • 316pp • H/C (ISBN: 9781522538530) • US $235.00

Design and Use of Virtualization Technology in Cloud Computing
Prashanta Kumar Das (Government Industrial Training Institute Dhansiri, India) and Ganesh Chandra Deka (Government of India, India)
Engineering Science Reference •©2018 • 315pp • H/C (ISBN: 9781522527855) • US $235.00

For an entire list of titles in this series, please visit:
https://www.igi-global.com/book-series/advances-computer-electrical-engineering/73675

IGI Global
DISSEMINATOR OF KNOWLEDGE

701 East Chocolate Avenue, Hershey, PA 17033, USA
Tel: 717-533-8845 x100 • Fax: 717-533-8661
E-Mail: cust@igi-global.com • www.igi-global.com

Table of Contents

Detailed Table of Contents

are abused to upgrade framework execution. At that point, big data-as-a-service, including big data infrastructure-as-a-Service, big data platform-as-a-service, and big data analytics software-as-a-service, is utilized to give regular huge information-related administrations (e.g., getting to benefit-produced huge information and information investigation results) to clients to improve effectiveness and lessen cost.

This chapter improves the SE scheme to grasp these contest difficulties. In the development, prototypical, hierarchical clustering technique is intended to lead additional search semantics with a supplementary feature of making the scheme to deal with the claim for reckless cipher text search in big-scale surroundings, such situations where there is a huge amount of data. Least relevance of threshold is considered for clustering the cloud document with hierarchical approach, and it divides the clusters into sub-clusters until the last cluster is reached. This method may affect the linear computational complexity versus the exponential growth of group of documents. To authenticate the validity for search, minimum hash sub tree is also implemented. This chapter focuses on fetching of cloud data of a subcontracted encrypted information deprived of loss of idea and of security and privacy by transmission attribute key to the information. In the next level, the typical is improved with a multilevel conviction privacy preserving scheme.

The purpose of this chapter is to observe the 3D asset development and product development process for creating real-world solutions using augmented and virtual reality technologies. To do this, the authors create simulative software solutions that can be used in assisting corporations with training activities. The method involves using augmented reality (AR) and virtual reality (VR) training tools to cut costs. By applying AR and VR technologies for training purposes, a cost reduction can be observed. The application of AR and VR technologies can help in using smartphones, high performance computers, head mounted displays (HMDs), and other such technologies to provide solutions via simulative environments. By implementing a good UX (user experience), the solutions can be seen to cause improvements in training, reduce on-site training risks and cut costs rapidly. By creating 3D simulations driven by engine mechanics, the applications for AR and VR technologies are vast ranging from purely computer science oriented applications such as data and process simulations to mechanical equipment and environmental simulations. This can help users further familiarize with potential scenarios.

Diabetes Mellitus has turned into a noteworthy general wellbeing issue in India. Most recent measurements on diabetes uncover that 63 million individuals in India are experiencing diabetes, and this figure is probably going to go up to 80 million by 2025. Given the rise of big data as a socio-technical phenomenon, there are various complications in analyzing big data and its related data handling issues. This chapter examines Hadoop, an open source structure that permits the disseminated handling for huge datasets on group of PCs and thus finally produces better results with the deployment of Iterative MapReduce. The goal of this chapter is to dissect and extricate the enhanced performance of data analysis in distributed environment. Iterative MapReduce (i-MapReduce) plays a major role in optimizing the analytics performance. Implementation is done on Cloudera Hadoop introduced on top of Hortonworks Data Platform (HDP) Sandbox.

This chapter presents an introduction to the basics in big data including architecture, modeling, and the tools used. Big data is a term that is used for serving the high volume of data that can be used as an alternative to RDBMS and the other analytical technologies such as OLAP. For every application there exist databases that contain the essential information. But the sizes of the databases vary in different applications and we need to store, extract, and modify these databases. In order to make it useful, we have to deal with it efficiently. This is the place that big data plays an important role. Big data exceeds the processing and the overall capacity of other traditional databases. In this chapter, the basic architecture, tools, modeling, and challenges are presented in each section.

Foreword

When I was invited to write a foreword for the book *Big Data Analytics for Satellite Image Processing and Remote Sensing,* I felt glad to note the varied tools, challenges, methods in Bigdata for Satellite Image processing. This book is a significant collection of 10 chapters covering image processing, satellite image processing, bigdata and cloud based processing, as well as their applications in emerged in the recent decades. This book provides an excellent platform to review various areas of Satellite Image processing and affords for the needs of both beginners to the field and seasoned researchers and practitioners. The tremendous growth of Satellite Image Processing and Bigdata are documented in this book, such as Bigdata, Satellite Image Processing, Remote Sensing, Computational methods, 3D Asset development and product development, Landslide susceptibility, Hierarchical clustering, Modified Support Vector Machine, Bigdata as a Service and Cloud Based Workflow Scheduling techniques which are focused in various applications.

To the best of my knowledge, this is the first attempt of its kind, providing a coverage of the key subjects in the fields Bigdata, Satellite Image Processing and Cloud Computing and applications. This book is an invaluable, topical, and timely source of knowledge in the field, which serves nicely as a major text book for several courses at both undergraduate, post graduate levels and scholars. It is also a key reference for scientists, professionals, and academicians, who are interested in new challenges, theories and practice of the specific areas mentioned above.

I am happy to commend the editors and authors on their accomplishment, and to inform the readers that they are looking at a major piece in the development of computational intelligence on organizational decision making. I am familiar with your research interests and expertise in the related research areas Big Data as a Service, Big Data Industry Standards, Experiences with Big Data Project Deployments. New Computational Models for image remote sensing and Big Data which would make an excellent addition to this publication. This book is a main step in this field's maturation and will serve to challenge the academic, research and scientific community in various significant ways.

S. Arunkumar Sangaiah
VIT University, India

Preface

As we have entered an era of high resolution earth observation, the Remote Sensing data are undergoing an explosive growth. The proliferation of data also gives rise to the increasing complexity of remotely sensed data, like the diversity and higher dimensionality characteristic of the data. This book aims to discuss and address the difficulties and challenges faced in handling big data and remotely sensed data applied in various applications. The editors have received chapters that address different aspects of using big data upon satellite image processing and related topics. Additionally, the book also explored the impact of such methodologies on the applications in which this advanced technology is being implemented.

This comprehensive and timely publication aims to be an essential reference source. This book helps the researchers who are working on satellite image processing with the available literature in the field of big data upon image processing techniques on remote sensing while providing for further research opportunities in this dynamic field. It is hoped that this text will provide the resources necessary technology to the developers and managers to adopt and implement these techniques, platforms and approaches in developing efficient solutions.

NEED FOR A BOOK ON THE PROPOSED TOPICS

Big data is an evolving term that describes any voluminous amount of structured, semi-structured, and unstructured data that has the potential to be mined for meaningful information. The scope of image processing and recognition has broadened due to the gap in scientific visualization. Thus, new imaging techniques have developed, and it is imperative to study this progression for optimal utilization. The data received from remote sensing satellite in various forms has been used to study various applications in real world by applying big data analytics.

Big Data Analytics for Satellite Image Processing and Remote Sensing is a critical scholarly resource that examines the challenges and difficulties of implementing big data in image processing for remote sensing and related areas. Featuring coverage on a broad range of topics, such as distributed computing, parallel processing, and spatial data, this book is geared towards scientists, professionals, researchers, and academicians seeking current research on the use of big data analytics in satellite image processing and remote sensing.

ORGANIZATION OF THE BOOK

The book is organized into 10 chapters. A brief description of each chapter is given as follows:

1. **New Computational Models for Image Remote Sensing and Big Data:** This chapter focuses on the development of new computational models for remote sensing applications with big data handling method using image data. Furthermore, this chapter presents overview of the process of developing systems for remote sensing and monitoring. The issues and challenges are presented to discuss various problems related to the handling of image big data in wireless sensor network that has various real-world applications. Moreover, the possible solutions and future recommendations to address the challenges have been presented and also this chapter includes discussion of emerging trends and conclusion.
2. **Big Data Computation Model for Landslide Risk Analysis Using Remote Sensing Data:** Nowadays, effective and efficient strategies have to acquire, manage and analyze data leads that to better decision making and competitive advantage. The development of cloud computing and the big data era, brings up challenges to traditional data mining algorithms. the characteristics of complexity have to formulate an extreme challenge for discovering useful knowledge from the big data. Spatial data is complex big data. To handle complex spatial landslide data, this chapter is proposed with Multi Ranking Decision Tree big data approach.. The Proposed Classifier performance is validated with massive real time dataset. The results indicate that our classifier exhibits both time efficiency and scalability.
3. **Modified Support Vector Machine Algorithm to Reduce Misclassification and Optimizing Time Complexity:** This chapter deals with optimization of the time consumption for computation of classification algorithm. All these results in reduction in miss-classification and faster computation of class for data point. This caused due to some drawback of each algorithm depending upon type of data. To reduce miss-classification new methodology of "Support Vector Machine" is to be introduced, instead of having hyperplane exactly in the middle position of hyper plane that has to be changed per number of data

points of class available near to the hyperplane. Thereby, this chapter shows some multi core architecture which has been used to compute more than one in dependent modules simultaneously.

4. **An Analysis of Usage-Induced Big Data:** This chapter deals with the "Analysis of Usage-induced Big Data". The most effective method to store, oversee, and make values from the administration situated enormous information turn into a vital research issue. With the inexorably huge measure of information, a solitary framework which gives normal usefulness to overseeing and dissecting diverse sorts of administration produced enormous information is critically required. This test is addressed in this chapter which gives a review of administration produced huge information and Big Data-as-a-Service that includes Big Data Infrastructure-as-a-Service, Big Data Platform-as-a-Service, and Big Data Analytics Software-as-a-Service, giving regular huge information related administration (e.g., getting to benefit produced huge information and information investigation results) to clients to improve effectiveness and lessen cost.
5. **Glorified Secure Search Over Encrypted Secured Cloud Database Through Hierarchical Clustering Computation: Search on Cloud Through Hierarchical Clustering:** This chapter proposed an improvised SE scheme to grasp the contest difficulties. In the development, prototypical, hierarchical clustering technique is intended to leader additional search semantics with a supplementary feature of making the scheme to deal with the claim for reckless cipher text search in big scale surroundings such situations where there is huge amount of data. This focus on fetching of cloud data of a subcontracted encrypted information deprived of loss of idea and if security and privacy by transmission attribute key to the information with the next level the typical is improved with the multilevel conviction privacy preserving scheme.
6. **Research Analysis of Development Pipelines in Augmented and Virtual Reality Technologies:** The objective of this chapter is 3D asset development and product development process and creation of real world solutions using Augmented and Virtual Reality Technologies. The methods involve creation of simulative software solutions that can be used in assisting corporations with training activities as using Augmented Reality (AR) and Virtual Reality (VR) training tools to cut costs. This chapter applied AR and VR technologies for training purposes, a 60% cost reduction can be observed. The application of AR and VR technologies can help in using a smartphones, high performance computers, head mounted displays (HMDs) other such technologies to provide solutions via simulative environments. Thereby the chapter proved improvements by implementing a good UX (user experience) in training reducing on-site training risks and cut costs rapidly.

7. **Iterative MapReduce: i-MapReduce on Medical Dataset Using Hadoop:** The goal of this chapter is to dissect and extricate the enhanced performance of Data Analysis in distributed environment. Iterative MapReduce (i-MapReduce) plays a major role in optimizing the analytics performance by finding useful data patterns using reduced mapper scans. The term 'Enormous Data' alludes to the monstrous volumes of both organized and unstructured information which can't be directly utilized with conventional database administration frameworks. With the quick increment of urbanization in India and its many-fold determinants, the information will become tremendous and turns out to be Big Data which couldn't be handled by traditional DBMS. In this chapter, Hadoop, an open source structure that permits the disseminated handling for huge datasets on group of PCs has been examined and thus finally produces better results with the deployment of Iterative MapReduce.
8. **Big Data for Satellite Image Processing: Analytics, Tools, Modeling, and Challenges:** This chapter presents an introduction to the basics in bigdata including architecture, modeling and the tools used. Bigdata is a term which is used for serving the high volume of data that can be used as an alternative to RDBMS and the other analytical technologies such as OLAP. In order to make it useful, the chapter has to deal it efficiently by placing the bigdata that exceeds the processing and the overall capacity of other traditional databases. This also discuss with the basic architecture, tools, modeling and challenges to give efficient solution using bigdata for satellite image processing.
9. **Big Data Processing: Application of Parallel Processing Technique to Big Data by Using MapReduce:** This chapter deals with a description about MapReduce which serves as a programming algorithm for distributed computing in a parallel manner on huge chunks of data which can easily execute on commodity servers. Thereby, reduce the costs of server maintenance and removal of requirement of having dedicated servers towards the running of processes. This chapter discuss about the various approaches towards map reduce programming model and how to use it in an efficient manner for scalable text based analysis in various domains like machine learning, data analytics and data science. Hence it will deal with various approaches of using map reduce in these fields and how to apply various techniques of MapReduce in these fields effectively and fitting the MapReduce programming model into any text mining application.
10. **Image Processing in CES-4019 Distinction Between Fake Note and a Real Note:** In the chapter, the author is trying to distinguish between fake note and a real note and would like to take it to a level where it can be used everywhere. The data after the detection of the fakeness and the real note can be stored in the database. As the data to store will be hug, to overcome this problem, the

chapter deals with Big Data. It will help us to store large amount of data in no time. The difference between real note and fake note is that real note has its thin strip to be more or less continuous while the fake strip has fragmented thin lines in the strip. One could say that the fake note has more than one line in the thin strip while the real note only has one line. Therefore, the chapter deals with: just one line, it's real but if the users' see more than one line it's fake. Here, the data-set used is foreign currency for the experimentation results.

Gopinath Ganapathy
Bharathidasan University, India

V. Susheela Devi
Indian Institute of Science Bangalore, India

Ravee Sundararajan
London South Bank University (LSBU), UK

Acknowledgment

It is obvious that the development of a book of this scope needs the support of many people. We must thank Dr. Marianne Caesar, Assistant Development Editor and the editorial team, IGI Global for their encouragement and support enabled the book publication project to materialize and contributed to its success. We especially thank the management of VIT University for their tremendous assistance. The most important contribution to the development of a book such as this comes from peer reviews. We cannot express our gratitude in words to the many reviewers who spent numerous hours reading the manuscript and providing us with helpful comments and ideas.

We would like to express our sincere gratitude to all the contributors, who have submitted their high-quality chapters, and to the experts for their supports in providing insightful review comments and suggestions on time.

Chapter 1
New Computational Models for Image Remote Sensing and Big Data

Dhanasekaran K. Pillai
Jain College of Engineering, India

ABSTRACT

This chapter focuses on the development of new computational models for remote sensing applications with big data handling method using image data. Furthermore, this chapter presents an overview of the process of developing systems for remote sensing and monitoring. The issues and challenges are presented to discuss various problems related to the handling of image big data in wireless sensor networks that have various real-world applications. Moreover, the possible solutions and future recommendations to address the challenges have been presented and also this chapter includes discussion of emerging trends and a conclusion.

INTRODUCTION

The goal of developing new computational models is to enable creation of new big data based remote sensing infrastructure for analysing and mining image data. The system must include a data collection component to aggregate, integrate data and perform validation of image data. Then, the central component of the system performs tasks like filtering, analysis and extraction of relevant patterns from image data. The result of extraction and prediction can be used for agricultural monitoring, crop monitoring or for forecasting of weather and market values.

DOI: 10.4018/978-1-5225-3643-7.ch001

Most of the big data framework that uses image remote sensing involves the following steps:

1. Organizing and integrating data from scenario based models, satellite images, and other remote stations.
2. Developing data mining and correlation analysis techniques to perform time-series data mining, spatial data analysis or spatiotemporal analysis.
3. Developing classification methods to perform image data classification.
4. Evaluating fitness of the data models by comparing data with standard values or indices.
5. Developing methods for monitoring activity using images with low or high resolution.

The system architectural model in Figure 1 involves scenario based models for analysing and mining images, weather data, and pollution data.

For data storage and management, DSpace can be used to store and maintain a large amount of heterogeneous data. The DSpace is an open source dynamic digital repository that can be used for image analysis while using big data. It enables free access to the data.

This chapter enables users to understand major issues and problems related to remote sensing in combination with big data handling for image data. After analysing solutions recommended for addressing the problems, users will be able to understand the process of developing a new framework, tools, or software systems to meet the current needs.

Figure 1. Computational system model using different types of data

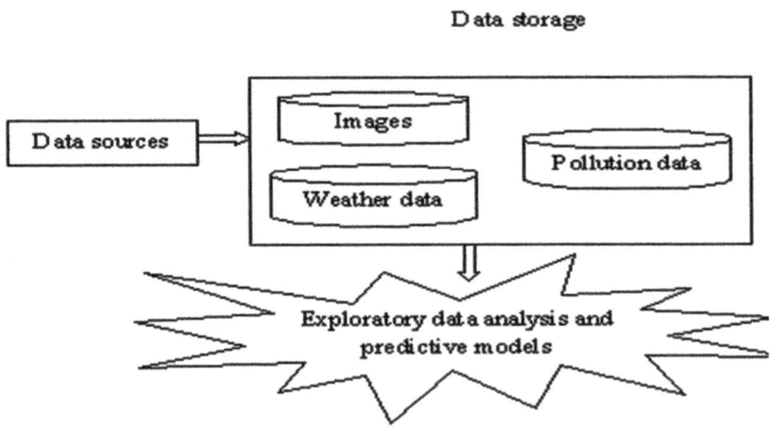

BACKGROUND

Mostly, remote sensing data is collected to analyse disease conditions, growth of plant, pollution, land use, road traffic congestions, and effects of disaster etc. One solution to address these problems is to develop possible computational models which represent several modules for the data analysis. The creation of thematic map for certain problems requires meaningful analysis which aims to show satisfactory results.

The analytical solutions to various engineering problems require robust analysis in a particular context. In this perspective, various issues related to scenarios should have to be addressed, because, complexity of different scenarios varies over time. For most of the existing developments, analysing and developing computational models have been the main motivation for remote sensing applications that use images.

The image data collected through multispectral image sensing can provide information at the element level. It can also provide information at the composite level via inter-pixel relationships. In some of the applications, the output information is used to assess the user belief or expert suggestions by the analyst. The software program validates the hypothesis developed by users. In most of the cases, the analysis fails due to the compatibility issues between user-defined performance measure that is used for optimization and the objective that is unlikely to produce the expected results. So, every analysis should need to be applied iteratively. The ordering and optimal selection of the objects involved in analysis may not be known. Hence, an effective approach must use suitable selection and ordering technique for objects in image remote sensing and analysis applications.

Mostly, practical engineering problems cannot be solved with hundred percent perfectness, because, the software system performance largely depends on the error-tolerance, resource availability, and time taken for the process. Engineering solutions require experimentation with iterative analysis of algorithms. The new computational models presented in this chapter will be helpful while proposing a new system to address various problems related to image remote sensing and big data.

Crop related mapping of soybean and corn has been conducted at regional scale focusing on the tropical and temperature plains (Arvor, Jonathan, Meirelles, Dubreuil, & Durieux, 2011). Most of the methods have used spectral features of land cover classes for classification either based on supervised learning methods or based on unsupervised learning methods.

Genetic Programming (GP) uses the principle of natural selection to discover information using software programs (Alavi & Gandomi, 2011). In fact, GP is a specialized version of genetic algorithms in which the encoded individual solutions are software programs rather than binary strings.

GP in some of the recent existing works has focused on the behaviour characterization and some of the studies have used GP as a tool for interpreting remote sensing data and can also be used to analyze ground movement patterns, object movement patterns, and change detection etc.

The crop classes may be identified from phonological information by deriving phonological metrics and by building classification rules based on crop calendar and stage-dependent crop conditions (Dong, Xiao, Kou, Qin, Zhang, & Li, 2015).

The effects of temperature, nutrients, and disease have been predicted by developing an efficient feature classification model (Kuttiyapillai & Rajeswari, 2014). This computational model was focused on information extraction considering tomato related feature analysis for classification using large margin k-nearest neighbour classifier.

A context-based method for extracting food safety information was discussed in a paper titled a method for extracting task-oriented information from biological text sources (Kuttiyapillai & Rajeswari, 2015). It was focused on information extraction and dynamic programming technique to find relevant genes from sequences.

In agricultural field, the detailed information of the planting area is required to improve the yield estimate, so, it can be incorporated into crop yield models (Lobell, Thau, Seifert, Engle, & Little, 2015). To find insights into planting area related and crop related information, image remote sensing that involves computational models would be a viable option.

Another approach based on machine learning has been developed to select and combine feature groups. It allows users to give positive and negative examples. This method improves the user interaction and the quality of queries (Minka, & Picard, 1997). The two methods discussed in the following paragraphs are based on the concepts of information mining.

Some of the existing approaches does not adapt to different situations to satisfy user needs. So, the image retrieval has been developed based on relevance feedback functions (Rui, Huang, Ortega, & Mehrota, 1998). Also, the system is designed to search image according to the suggestions of user, taking the feedback into account.

During the last decades, traditional database system has been used to store data considering characteristics such as color, texture, and shape. Further development has focused on region-based image retrieval since the content-based image retrieval was not satisfactory due to the growing size of image and information content (Veltkamp, Burkhardt, & Kriegel, 2001). This method has been found to be a viable solution to deal with the varying nature of image content. In this method, each image is segmented, and object characteristics are used to index individual object.

Due to the atmospheric effects such as Rayleigh scattering that occurs because of atmospheric molecules, ozone, absorption by water vapour, and other gases, absorption due to atmospheric aerosols, changes will frequently occur in the data collection environment. So, the data correction becomes a computationally intensive task which requires innovative error correction approaches, for example, standard radiative transfer algorithms 6S (Vermote, Tanre, Deuze, Herman, & Morcrette, 1997) for processing high resolution image datasets. Otherwise, the data processing will not be practically feasible.

MAIN FOCUS OF THE CHAPTER

The advances in remote sensing technologies and availability of Internet access have increased the amount of data gathered from different locations via satellite imaging technique. The remote sensing data gathered by a single data center of satellite are dramatically increasing at several terabytes rate per day. Recent study shows that the data gathered would even exceed Exabyte range because of the emergence of high-resolution earth observational data which has led to the curse of dimensionality issues in terms of image data. These data in some sense termed as big data.

The way earth system changes would change the accuracy and updation of current global earth data. To handle the increasing complexity in terms of size and dynamism, multi-sensor remote sensing data and temporal big data mining are useful for big data processing. The major data-intensive computing issues and challenges arise because of rapid growth of remote sensing data. So, the methods to deal with the computational complexity associated with big data are necessarily moving towards the high-performance computing paradigm.

Because of the data availability requirement and huge required computing power for processing massive amount of data, the cluster-based high-performance computing still remains as big challenge in remote sensing applications. Moreover, the intensive irregular data access patterns of remote sensing data increase I/O burden making the common parallel file systems inapplicable. Further, the current task scheduling seeks load balancing among computational resources taking the data availability as major concern. To reduce the complexity, the large tasks could be divided into smaller data dependent tasks with ordering constraints. Another way to handling the critical issue is to introduce an optimization technique for scheduling of tasks while trying to achieve higher performance.

In addition, specifically the Message Passing Interface (MPI) enabled cluster systems that support multilevel hierarchy with increasing scale, lead to cause difficulty and error-prone tasks while using parallel programming techniques for remote sensing applications. In the near future, the increasing demand for real-time processing would further increase the issues associated with remote sensing applications. However, unlike traditional systems, the emerging remote sensing systems can provide timely real-time processing of sensed data gathered from environment.

Remote sensing is defined as the technology for perceiving an object or surface from a remote distance to identify the characteristics of an object. It requires data acquisition instruments or devices. Remote sensing data are treated as big data because big data is not only refers to the volume and velocity of data which causes computational complexity, other issues such as variety, complexity (or high dimensionality) and trustworthiness also makes these data as big data.

Normally, the processing of remote sensing big data involves the following stages on the process flow: satellite observational network, data acquisition and recording, remote sensing data processing (pre-processing, central processing, information abstraction and representation), and operating remote sensing applications.

The pre-processing techniques are used to remove noise and to correct inconsistent data. It involves techniques like smoothing, outlier detection, and dimensionality reduction and so on. The central processing component is responsible for further correction, modification and merging of relevant data. The information abstraction and representation would involve machine learning techniques to perform efficient tasks like classification, clustering, and association analysis.

Some of the examples of remote sensing and monitoring include: disaster response and monitoring, climate change control and monitoring, water management and usage monitoring, industry operation control and monitoring, fraud detection and monitoring etc. Example sensor support includes spatial sensors, temporal sensors, temperature sensors, humidity sensor, moisture sensor etc. Example data on remote sensing and monitoring include GPS data, GIS data, and other sensors' data. The processing of massive remote sensing data poses challenging issues; some of these issues are:

1. Difficulty in managing remote sensing big data.

Naturally, the remote sensing data captured from different data centres are distributed and these data centres are normally far away and connected by the Internet. So, the data management component has to critically manage these distributed, huge amounts of data for improving interoperability and global data sharing.

This means that there is a demand for introducing more storage devices and for improving easy access. In order to tackle these challenges, new technologies and techniques are required. Further, the high dimensionality characteristics of remotely sensed data makes the distributed data sharing and accessing more complicated. The main issue arises while trying to organize and map the multi-dimensional remote sensing imageries to a one-dimensional array.

In addition, the remote sensing data partition requires suitable space-filling curves and efficient data organization depends upon data availability. Moreover, the complex metadata in structured form associated with the remote sensing data makes data processing little more complicated. It also affects the efficiency on data storing and indexing of metadata.

2. The irregular data access pattern.

The critical data processing and data sharing component requires high data availability. Most of the remote sensing applications perform processing on irregular data access pattern. Example issues include: irregular I/O patterns and increased CPU load that occurs by varying degree of dependency between computation of algorithm and remote sensing data.

3. The data loading and transmission of big data.

Because of the high dimensionality of the massive amount of remotely sensed big data, data loading, memory management, and data transmission becomes little more complicated and inefficient. The capacity of image big data may sometimes go beyond the limited capacity of computer memory when multi-dimensional images are considered along with complex metadata.

Hence, it requires large, efficient data structure to store these massive amounts of remotely sensed big data in local memory. Further, data transmission among processing nodes requires high bandwidth to transmit image big data in the form of data blocks, so, it would be a time-consuming process when the volume of data to be communicated is large.

4. Scheduling of large number of data-dependent tasks.

The large scale design modeling of water management and remote monitoring would involve a large number of smaller data dependent tasks. Therefore, the main processing module become extremely difficult and may require ordering constraints to deal with data dependent tasks. To achieve good performance, an optimized scheduling algorithm is required. Sometimes, decoupling of data dependencies may be helpful to achieve better selection of execution path.

5. Efficient parallel programming.

The conventional clustering systems must deal with a multilevel hierarchical organization and rapidly increasing scalability issues. It requires efficient parallel programming techniques for dealing with data intensive computing tasks in remote sensing and real-time monitoring applications. Recently, OpenMP has been designed as a new paradigm for shared-memory cluster computing.

SOLUTIONS AND RECOMMENDATIONS

Solutions and recommendations in dealing with the issues, controversies, or problems are presented in this section. To address various issues and problems related to remote image sensing, this chapter presents solutions in the form of computational models and recommends some problem solving techniques.

1. High-Performance Computing Model

The remote sensing and monitoring that requires complex models to work with large amounts of multi-sensor and temporal data sets may have to apply widely used pre-processing algorithm. The computational and storage requirements for problems that use large number of earth observations would normally exceed the available computing power on a single computing platform.

To deal with this computational complexity, a new computational model is introduced combining scalable and distributed heterogeneous network with high performance software program for huge data processing, indexing, and organizing remote sensing data. The high-performance computing model in Figure 2 uses a hierarchical data management to achieve load balancing among the logical sensor processing nodes and also focuses on minimizing I/O overheads.

This computational model deals with error correction on remotely sensed image data, and includes major components such as image segmentation using hierarchically connected components, followed by retrieval of data distribution using computing function, hierarchy based data organization that allows on-demand processing and retrieval of information. This scheme with addition or removal of some software modules can be adopted to improve computational time in remote sensing based applications.

Most of the information analytical systems require data fusion from various sources and instruments. The use of these data sets in modeling ecosystem response, various types of data face challenges in dealing with huge data storage and high

Figure 2. High performance computing model for remote image sensing

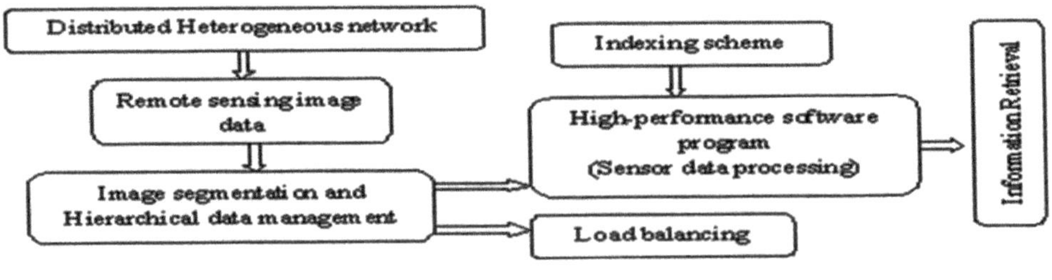

computational complexity. Therefore, the data acquisition, processing, mapping and conversion of remote sensing data deal with a complicated modeling and increased computational complexity. The computational tasks that involve a variety of pre-processing of satellite data (e.g. land use data) include complex neighbourhood operations.

For example, the total storage requirements for a global land cover data sets would sometimes require processing of Tera bytes, Giga Floating Point Operations (GFLOPs), or Peta FLOPs for data processing. So, it requires high performance computing techniques to acquire information that is required from earth observational systems.

Although innovative parallel processing algorithms have been developed for processing of large datasets in a heterogeneous, distributed environment, still there are promising avenues for further improvements and new developments. Normally, remote image processing includes steps such as pre-processing, image segmentation, retrieval of distribution via function, and software module for processing, storing and retrieval of data.

2. Cloud Based Big Data Processing Model

Finding task-relevant information has become a tedious task because of massive volume of real-time data. To extract meaningful information, the system requires efficient data analysis, aggregation and storage mechanism while collecting remote data. The cloud based image big data processing model in Figure 3 consists of four major units such as Data Acquisition unit for remote sensing data (DA), Local Pre-processing (LP), Local Storage and Processing (LSP), and Cloud Storage and Retrieval (CSR).

Cloud data processing has capability to handle scalability issue while dealing with remotely sensed big data. The system that uses earth observational system is required

Figure 3. Cloud based remotely sensed image big data processing model

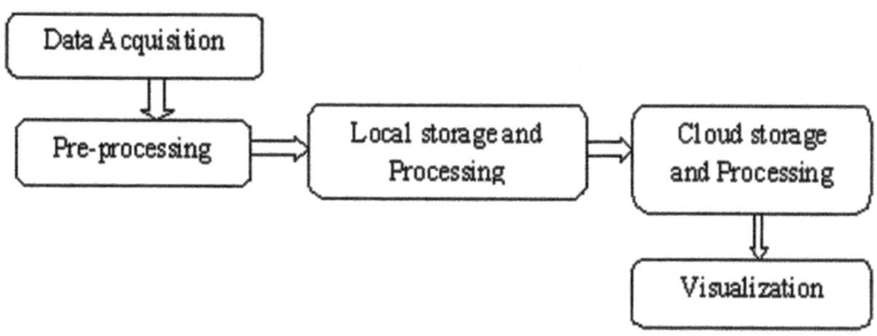

to perform local processing to purify raw data which usually has inconsistent data. Because of size and complexity of big data, the traditional database management faces difficulties in handling big data.

The major challenges in big data processing include: 1) volume that denotes large amounts of data generated from remote area; 2) velocity that denotes frequency and speed at which data has been generated and shared; 3) variety that denotes diversity of data types of data collected from various sources. A system that deals with any of these challenges may use smaller subsets to create a result set through correlation analysis. The major disadvantages of conventional systems include: 1) difficult data transformation of remotely sensed continuous stream of data; 2) data collected from remote areas are not in a valid format which is ready for further data analysis; 3) remote sensor network may generate vast amounts of raw data.

To deal with various challenges in remotely sensed big data analysis, a system that incorporates offline data storage and filtering with load balancing sub-systems can be developed for extracting useful information. The input to the big data system comes from social networks, satellite imagery, sensor devices, Web servers, finance data store, and banking data store etc.

In this computational model, the load balancer balances the processing power by distributing the real-time data to the servers where the base station is processing data. It can also enhance the efficiency of the system. Data extraction finds insights into the data model and discovers information to create a structured view of the data. Here, machine learning techniques are applied to process and interpret image data for generating maps, and summary results. The system that deals with huge amounts of big data processing can be implemented in development platforms like Python, R Analytics platform, and Hadoop using MapReduce.

Parallel Processing Model Using I/O Optimization

The data-intensive application that provides on-demand information requires efficient indexing schemes and data replacement techniques to achieve the maximum parallel I/O throughput. These techniques are widely used for handling spatial and temporal datasets with different resolutions.

The atmospheric effects will vary according to the context based on spatial and temporal data, and also depends on the wavelength and geometry of observations. Sometimes decoupling of the effects of components will be useful in remote sensing applications. The feature selection based error removal will also be helpful in dealing with erroneous data.

The parallel algorithm based on connected components can efficiently perform while dealing with complicated, computationally intensive tasks like image enhancement and segmentation for classification of regions in remotely sensed images. This means that this type of algorithm can improve accuracy and processing time.

The error correction method involves two steps. The first step is to estimate atmospheric properties from the imagery. The second is for retrieving surface reflectance. The following steps are involved in this method:

- Input image (i.e. Window of pixels)
- Identifying dark targets (i.e. dense green vegetation)
- Setting the reflectance threshold
- Compute mean value of pixels whose reflectance threshold is less than the set threshold
- Estimate reflectance in Thematic Mappers (TM) for the dark pixels
- Use a lookup table to estimate the aerosol optical thickness in Thematic Mapper bands. Then, check whether the estimate is larger than or equal to the set aerosol optical thickness. Otherwise, repeat the previous steps with a smaller threshold until the first aerosol optical thickness is greater than or equal to the third aerosol optical thickness.
- Use the exponential relationship between wavelength and aerosol optical thickness to derive coefficients
- Once the aerosol optical thickness is determined for all the TM bands, use the lookup table to apply error correction in the central pixel of the window.
- The window is moved by incrementing pixel position until entire image is covered through repeated execution.

The process steps involved in parallel processing model for image remote sensing is shown in Figure 4.

In this computational model, the correlation between estimated measurements and ground or standard measurements are verified. The resultant image after applying error correction method looks clearer than the original image. Also, the spatial pattern of aerosol optical thickness maintains consistency in comparison with the normal image. The experimental setting for exchanging messages simultaneously among all the processor or processing nodes have to connect all the processors by an Ethernet and a high-performance switch. The approach that uses single program multiple data model can be suitable to run the same code on different parts of the input image in parallel.

To implement parallel processing, the input image is divided into a number of blocks. Then, the equal-sized blocks are distributed among all the processing nodes. The linear running time is measured in terms of a function of the number of nodes. This computational model can be useful to minimize the running time by an efficient error correction method. Both qualitative and quantitative analysis is used to improve

Figure 4. The parallel processing model for image remote sensing

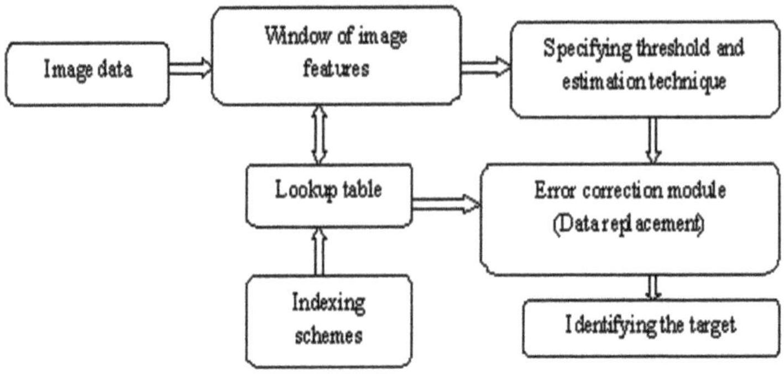

reliability of the computational model. The modular approach of implementation of individual components allows easy modification to the atmospheric properties. The segmentation of remotely sensed imagery is used to form clusters which contain similar type of region consisting of pixels. These regions can be classified into categories for improving prediction accuracy.

Information Mining Model in Remote Sensing Images

The content-based information mining system focuses on managing and exploring large volumes of remotely sensed image data. This type of system may include offline processing as well as an interface to online extraction and processing. The offline module extracts image features, and attempts to improve compression, data reduction, generation of index for unsupervised image, and storing content in the database. Further, the semantic interpretations of the image content (i.e. user's interest) are associated with Bayesian network based on index of content. Since this method obtains only a few training samples, the link computation can be done via online searching on the complete image repository. The Figure 5 illustrates information mining model for remote sensing images.

During the past few decades, the image satellite sensors such as optical sensors, synthetic aperture radar, and other satellite sensors have acquired huge amounts of image scenes. In future, the quantities of real-time image sensing data will further increase due to the data collection by high resolution satellite sensors. The state-of-the-art systems access these data and images through query passed using geographical coordinates, time of acquisition, and sensor type. The information which are collected using traditional system is often less relevant, so, only a few images can be used. Because the relevancy is determined by the content of the image considering its structures, patterns, objects, or scattering properties.

Figure 5. Illustration of information mining model for remote sensing images

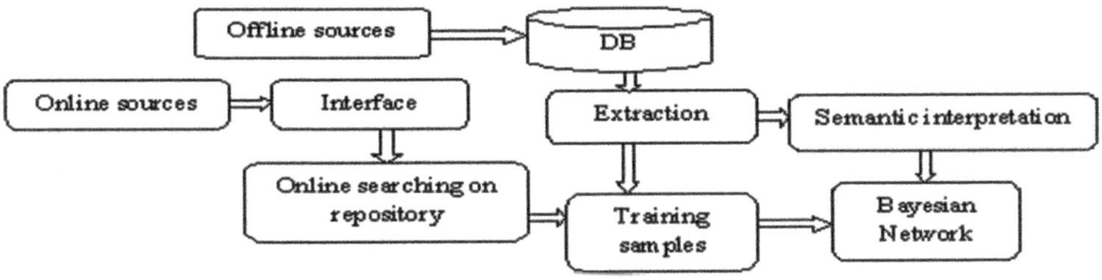

In this approach, a system which incorporates the traditional database concept for offline images and online search with semantic interpretation are recommended to develop efficient information mining system that applies Bayesian network for efficient reasoning. This system can make decision according to the interest of users. Based on the semantic interpretation of image content that is linked with index of content in Bayesian network, the user can search for relevant images which are stored in online repository. Then, the classification of images is performed to represent image contents that supports predicted target.

Computational Method for Analyzing and Mining Images

Remote sensing plays a very important role in delivering real-time information on the crop area for improving productivity, and environment. Traditional mapping of image patterns is difficult due to the high cost of repeated training data, inconsistency in interpretation, and the difficulty of handling the varying weather and crop growth.

In this section, an automated approach to map hybrid tomato and original tomato is discussed for analyzing and mining image patterns. Here, a decision tree classifier is constructed by using rules that are manually written based on expert opinions. The automated approach will be more advantageous when mapping is to be done for multiple years, because, the mapping can be performed without re-training or repeated calibration task. To identify vegetation, moderate resolution, time series based, imaging Spectroradiometer and reflectance product can be used which identifies hybrid variety of tomato and original variety based on year.

The surface reflectance of the shortwave infrared band is scaled to identify similarity between phenology of the two crops. The mapped areas can be verified with the standard statistical values that are maintained at the municipal level. The effect of mixed pixel can also be identified by applying classification technique and the reference dataset can be used to evaluate the resultant map. Furthermore, the automated mapping can be applied to other image series in future considering different scales and high-resolution images.

In this model, the following variables are used for measurement and classification:

V_b: This denotes the Enhanced Vegetation index value of background in the non-growing season.
V_a: This denotes the amplitude of Enhanced Vegetation index variation within the growing cycle. The V_a value of field crops is higher than natural vegetation. Average V_a of hybrid tomato is greater than original crop.
p, q: changing rate parameters that corresponds to the increased and decreased segments in the cycle. In this case, the field crop cycles have fast increase or decrease in enhanced vegetation index value.
D_i, D_d: These variables denote the middle Dates of segments when increasing or decreasing rates are high. These variables are used as indicators of the dates on which rapid growth is there and harvesting is necessary.
D_1, D_2, D_3, D_4: These variables denote when the second derivative of the curve reaches local maximum or minimum. D_1 denotes starting dates and D_4 denotes end dates of the growing season.
L: This variable denotes the difference between D_4 and D_1 . It represents the length of growing season. The length should maintain consistency. Hybrid tomato has shorter growing time than original tomato.
R: This variable denotes reflectance at D_i. Reflectance of Hybrid variety is slightly higher than normal variety. Here, pixels with high reflectance are hybrid, and pixels with low reflectance are normal variety and this provides confidence pixels for training.

Here, the automated mapping is performed based on classification rules in a tree-like structure. The variables which represent the changes in the vegetation and stages of transition during the growth period are useful in classification of season based crop cycles. The Figure 6 illustrates the data and control flow representation of mapping crop image patterns.

If there is high variability on seasons, less restrictive rules can be used to improve the repeatability of algorithm under different conditions. Further, additional classification rules are necessary based on pheno-spectral variables that utilize spectral properties at different stages for improving the separation.

Machine Learning Methods for Remote Sensing

Machine learning is a subfield of artificial intelligence. With an intention to develop algorithm that learns from machine readable data, machine learning has been evolved as a key technique in many application domains such as data mining,

Figure 6. Illustration of the data and control flow representation of mapping crop image patterns

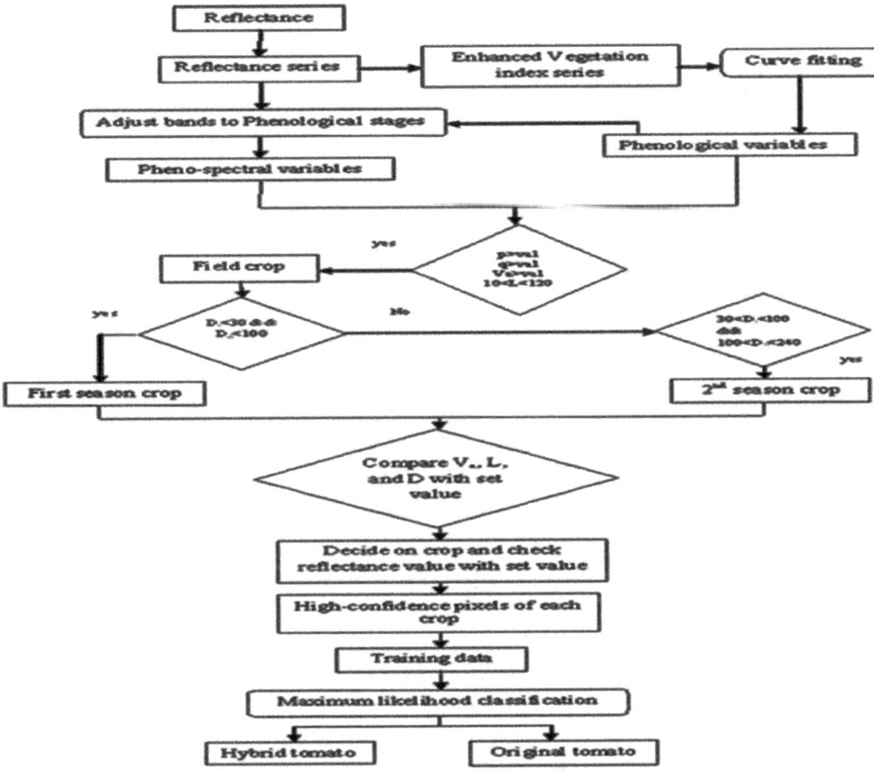

telecommunication, market analysis, and other software applications. It includes algorithms such as decision tree, Bayesian reasoning, Naive Bayes, k-nearest neighbour, support vector machine, neural networks, self-organizing map, and ensemble methods such as random forests, genetic algorithm, case-based reasoning, genetic programming, fuzzy logic, neuro-fuzzy, etc. The main motivation to apply machine learning technique for image remote sensing is to provide multivariate, nonlinear, and non-parametric regression or classification. It plays crucial role in science and engineering. Machine learning is capable of tackling various problems associated with geosciences and remote sensing.

Machine learning has two categories of methods, namely, supervised, unsupervised, which is used to develop a regression approach or classification approach of nonlinear systems. The system that involves this technique would involve a few or literally thousands of variables. In machine learning, training dataset consists of pre-classified vectors of image data. A subset of the training samples may not require separate validation. If theoretical knowledge is incomplete, and there are some observations and data, then, machine learning would be helpful to address such type of problems.

In the past few decades, machine learning has been a widely used technique in earth science and observational systems such as land use detection, ocean change detection, road extraction, and atmosphere, disaster prediction, crop disease prediction, activity detection, etc. Herein, a number of relevant applications of machine learning are summarized for understanding its applicability in geosciences and remote sensing. The main focus may fall under two categories, one is on how to apply multivariate nonlinear nonparametric regression, and the other is on how to use multivariate nonlinear unsupervised classification.

The machine learning is actually a universal approximation technique that learns the underlying behaviour pattern from a set of training data. Another fact about this technique is that, it does not require prior knowledge about the nature of the relationships among the data. The application of machine learning includes five areas: 1) deterministic model which is computationally expensive; 2) non-deterministic model, or an empirical model that can be derived from the existing data; 3) classification model; 4) clustering model and; 5) hybrid model. The study of the successful application of fuzzy logic, neuro-fuzzy, fuzzy-genetic, neuro-computing can be found in the fields such as oil exploration, intelligent reservoir exploration, crop area mapping etc.

Among all other machine learning methods, application of genetic programming (GP) in the field of remote sensing and image analysis has been restricted to a very few areas and is new also. Although techniques like support vector machine (SVM) and artificial neural networks (ANNs) have shown good performance, these machine learning follows black-box models. That is, they are not capable of producing practical prediction. Hence, GP is considered as a viable technique to deal with practical issues in remote sensing applications.

The variants of GP, namely, multi expression programming (MEP), linear genetic programming (LGP), and gene expression programming (GEP) are used to predict the uniaxial compressive strength and tensile strength of chalky and clayey soft limestone. GP models give high prediction than other existing models, which have used parameters as predictor variables, for example, modulus of elasticity of intact rock, uniaxial compressive strength, the number of joint per meter, rock mass quality designation, dry density, and geological strength index. In rock structure designing, GP can estimate uniaxial compressive strength of rocks that can be formulated in terms of effective water absorption by weight, porosity, and actual weight.

Some example applications of GP in remote sensing and big data analysis include estimation of the heavy rainfall near disaster-prone area which uses multi-variable meteorological satellite data, monitoring water quality using images, mapping of chemical contaminants in food items, landslide detection using image threshold, mapping crop planting area and soil moisture analysis. The major processes in image remote sensing involve image characterization and image classification.

Image classification is the process of assigning images into a predefined target classes according to the characteristics of images. It includes techniques like pre-processing, feature extraction, object detection, object segmentation, and object classification. Image classification is a challenging task in application domains such as video surveillance, biomedical imaging, vehicle traffic detection and navigation, industrial monitoring, remote sensing and real-time monitoring.

Some of the popular machine learning algorithms that can be used in image classification includes Support Vector Machine, k-nearest neighbour algorithm, Artificial Neural Network, Genetic Algorithm, Expectation Maximization Algorithm, C4.5 decision tree algorithm, AdaBoost, CART, k-means algorithm etc. Mostly, the machine learning algorithms are used to predict the future trends or quality decisions. Data mining task which involves machine learning aims to discover information and generate a large number of rules. In big data mining, image classification mainly focuses on classifying new objects or unknown vectors under a predefined category or target label.

Image mining is one of the domains which can be used to extract meaningful image content from a large image dataset. Image classification automatically classifies image pixels into appropriate target based on natural evaluation and relationships among image data. It has two categories of classification, namely, supervised classification, and unsupervised classification.

In disaster response domain, aerial imagery is captured through unmanned vehicles within few hours whereas satellite imagery can be captured in days. Also, the spatial resolution of the aerial imagery is higher than the imagery generated by satellites. To process large volumes of aerial image data in real-time, machine learning is applied to provide solution. Crowd sourcing provides a way to annotate features of interest such as damaged shelters, blocked roads by debris, collapsed building etc. These features of aerial images have been used to train a supervised machine learning systems which learns to interpret features from unknown or new images. In addition, the disaster response system includes a text processing module that can be used to send messages to the rescue team. Another approach to solve this problem is to develop a hybrid system which deals with both aerial images as well as satellite images.

FUTURE RESEARCH DIRECTIONS

Remote sensing of image big data for analysis and interpretation has several special characteristics such as dynamic behaviour, multi-scale, multi-source, high-dimensional, isomer, and non-linear characteristics. While analysing which are the characteristics is closely related to the data acquisition, it is observed that the intrinsic characteristics include dynamic-state, multi-scale, and non-linear characteristics. However, other characteristics are extrinsic characteristics for remote sensing big data.

Remote sensing provides a method to quickly and directly acquire data from earth surface. Emerging trends in remote sensing of big data in information science and environmental engineering has led to the application of remote sensing and monitoring techniques in various fields which include ecology, earth quake prediction and analysis, soil contamination, air pollution analysis, water pollution analysis, environmental geology, solid waste detection and monitoring, street light monitoring, crop disease analysis and loss prediction, industrial fraud detection and monitoring, weather forecasting, customer behaviour prediction, patient monitoring, home appliances monitoring, gas leakage detection and monitoring in industry, energy consumption and sustainable development etc. The computational model discussed in this chapter shows the importance of a particular model in a problem domain.

In recent years, major countries have launched remote sensing satellites, which include India, USA, and Russia. The remotely sensed features and data may differ based on image resolution, spectrum, mode of imaging, and revisit cycle, amplitude, and time. Nowadays, there are different remote sensing systems. Example low-resolution satellite imaging includes meteorological satellite MODIS, and microwave satellite Envisat. Example mid-resolution satellite imaging includes terrestrial satellite (e.g. Landsat), satellite with long revisit period (e.g. EO-1), microwave satellites (e.g. Terra, and RADARSAT). Example high-resolution satellite imaging includes QuickBird, IKONOS, and WorldView. It requires efficient investigations and techniques to deal with increasing diversity of data.

The satellites can also be classified according to the revisit cycle in an observational area. For example, Geostationary Operational Environmental Satellite (GOES) can provide high-quality, continuous stream of time correlated observational data from Earth surface. Example satellites that use short revisit cycle include MODIS, WorldView and RapidEye.

Another challenge in handling remote sensing data is to deal with increasing size or large volume of data. In image remote sensing, for a single scene, the volume of data may be at the gigabyte level or at the terabyte level. The satellite remote sensing data collected for a particular period in one country is maintained as historical data. The volume of these data may be at the petabyte level and a large archive maintained at the global level may reach up to Exabyte level. Therefore, remote sensing data is termed as “big data” and requires efficient big data handling techniques.

CONCLUSION

This chapter presented different computational models for image remote sensing and big data handling. The representation of remotely sensed image information on a hierarchical form or in suitable form with different semantic abstraction is based on the levels involved in computational models. For example, a Bayesian model may consist of the following levels: 1) extracting image features and meta-features using signal models; 2) obtaining a vocabulary of signal classes for each model by applying unsupervised machine learning (or clustering) of the pre-extracted image parameters; 3) At last, user-interests, i.e., semantic labels are linked to combinations of these vocabularies through Bayesian networks. In order to infer information from the image data that covers the class label or target label, the system has to learn the probabilistic link based on user given input samples.

Automatic extraction of meaningful information from remote sensing images involves information mining techniques and robustness evaluation on the extracted information from the observed image data. This model focused on extracting structural information from images by selecting prior models that correctly explains the structures within an image. On the lowest level, stochastic models are applied to capture spatial, spectral, and geometrical structure information from image. In recent years, this type of parametric models has been found to be suitable to characterize spatial information in images. High-order models are mostly used to deal with complex structures that have features at different scales (e.g. mountains, rivers, etc.). When neighbourhood size increases, the number of parameters increases and has to face averaging effect of different parameters. As a result, less discriminative power of the extracted features may cause ineffective interpretation of data.

To overcome this drawback, a multi-resolution image data cube that has the original image at the lowest layer and reduced resolution representations of the image at the higher layers of data cube may be generated. By applying texture model that uses Gibbs random field to layers of limited neighbourhood size, various information from different structures can be extracted. It provides a way to characterize a large set of spatial information. The output of feature extraction may have large volumes of data, which may be difficult to store and manage in practical applications. Moreover, clustering may reduce the accuracy due to a large data reduction. To remove unnecessary structures and to avoid the time-consuming process of similarity checking, clustering is performed across all images. Even if a model generates large number of clusters, the algorithm can show good efficiency by applying some of these computational models that focused on parallel processing to improve efficiency of the system to produce promising result.

REFERENCES

Alavi, A. H., & Gandomi, A. H. (2011). A robust data mining approach for formulation of geotechnical engineering systems. *Engineering Computations*, *28*(3), 242–274. doi:10.1108/02644401111118132

Arvor, D., Jonathan, M., Meirelles, M. S. P., Dubreuil, V., & Durieux, L. (2011). Classification of MODIS EVI time series for crop mapping in the state of MatoGrosso. *Brazil International Journal of Remote Sensing*, *32*(22), 7847–7871. doi:10.1080/01431161.2010.531783

Dong, J., Xiao, X., Kou, W., Qin, Y., Zhang, G., Li, L., ... Moore, B. III. (2015). Tracking the dynamics of paddy rice planting area in 1986-2010 through time series Landsat images and phenology-based algorithms. *Remote Sensing of Environment*, *160*, 99–113. doi:10.1016/j.rse.2015.01.004

Kuttiyapillai, D., & Rajeswari, R. (2015). A method for extracting task-oriented information from biological text sources. *International Journal of Data Mining and Bioinformatics*, *12*(4), 387–399. doi:10.1504/IJDMB.2015.070072 PMID:26510293

Kuttiyapillai, D., & Ramachandran, R. 2014. Design and analysis of feature classification model using information extraction in tomato growing environment. International Information Institute (Tokyo). Information, 17(8), 3947-3959.

Lobell, D. B., Thau, D., Seifert, C., Engle, E., & Little, B. (2015). A scalable satellite-based crop yield mapper. *Remote Sensing of Environment*, *164*, 324–333. doi:10.1016/j.rse.2015.04.021

Minka, T. P., & Picard, R. W. (1997). Interactive learning using a "society of models". *Pattern Recognition*, *30*(4), 565–581. doi:10.1016/S0031-3203(96)00113-6

Rui, Y., Huang, T. S., Ortega, M., & Mehrota, S. (1998). Relevance feedback: A power tool for interactive content-based image retrieval. *IEEE Transaction on Circuits System and Video Technnology*, *8*(5), 644–655. doi:10.1109/76.718510

Veltkamp, C. R., Burkhardt, H., & Kriegel, H. P. (2001). *State-of-the-Art in content-based image and video retrieval*. Norwell, MA: Kluwer. doi:10.1007/978-94-015-9664-0

Vermote, E. F., Tanre, D., Deuze, J. L., Herman, M., & Morcrette, J. (1997). Second simulation of the satellite signal in the solar spectrum, 6S: An overview. *IEEE Transactions on Geoscience and Remote Sensing*, *35*(3), 675–686. doi:10.1109/36.581987

KEY TERMS AND DEFINITIONS

Big Data: The large volume of data, or the complex data that are normally difficult to handle through simple computational model and techniques.

Computational Model: An abstract representation of the system that performs computations based on an approach developed by developers or scientists to solve computational problems.

Data Mining: The process of extracting meaningful information from datasets that are generated from data sources.

Image Classification: The process of classifying unknown data or new patterns or new images according to the existing, pre-classified vectors.

Machine Learning: The process of using the existing knowledge and considering the feedback as the system changes the state to discover required information in order to produce good results.

Parallel Processing: The process of performing multiple tasks simultaneously, reducing the waiting time of the process.

Remote Sensing: The task of identifying images and structural characteristics.

Satellite Imaging: The process of using the images from environment, to make the images in a system-dependent format.

Chapter 2
Big Data Computation Model for Landslide Risk Analysis Using Remote Sensing Data

Venkatesan M.
National Institute of Technology Karanataka, India

Prabhavathy P.
VIT University, India

ABSTRACT

Effective and efficient strategies to acquire, manage, and analyze data leads to better decision making and competitive advantage. The development of cloud computing and the big data era brings up challenges to traditional data mining algorithms. The processing capacity, architecture, and algorithms of traditional database systems are not coping with big data analysis. Big data are now rapidly growing in all science and engineering domains, including biological, biomedical sciences, and disaster management. The characteristics of complexity formulate an extreme challenge for discovering useful knowledge from the big data. Spatial data is complex big data. The aim of this chapter is to propose a multi-ranking decision tree big data approach to handle complex spatial landslide data. The proposed classifier performance is validated with massive real-time dataset. The results indicate that the classifier exhibits both time efficiency and scalability.

DOI: 10.4018/978-1-5225-3643-7.ch002

INTRODUCTION

Very large amount of Geo-spatial data leads to definition of complex relationship, which creates challenges in today data mining research. Current scientific advancement has led to a flood of data from distinctive domains such as healthcare and scientific sensors, user-generated data, Internet and disaster management. Big data is data that exceeds the processing capacity of conventional database systems. The data is too big, moves too fast, or doesn't fit the strictures of your database architectures. For instance, big data is commonly unstructured and require more real-time analysis. This development calls forms system architectures for data acquisition, transmission, storage, and large-scale data processing mechanisms. Hadoop is a platform for distributing computing problems across a number of servers. First developed and released as open source by Yahoo, it implements the MapReduce approach pioneered by Google in compiling its search indexes. Hadoop's MapReduce involves distributing a dataset among multiple servers and operating on the data: the "map" stage. The partial results are then recombined: the "reduce" stage. To store data, Hadoop utilizes its own distributed file system, HDFS, which makes data available to multiple computing nodes.

Natural disasters like hurricanes, earthquakes, erosion, tsunamis and landslides cause countless deaths and fearsome damage to infrastructure and the environment. Landslide is the one of the major problem in hilly areas. Landslide Risk can be identified using different methods based on the GIS technology. In Ooty, Nilgiri district, landslide was happened due to the heavy rainfall and frequent modification of land use features. Landslide disaster could have been reduced, if more had been known about forecasting and mitigation. So far, few attempts have been made to predict these landslides or prevent the damages caused by them. In the previous studies, various approaches were applied to such problems which show that it is difficult to understand and tricky to predict accurately. In order to analyze these landslides, various factors, such as Rainfall, Geology, Slope, land-use/land cover, soil and Geomorphology are considered and the relevant thematic layers are prepared in GIS for landslide susceptibility mapping. The data collected from various research institutes related to land slide helped to predict and analyze the land slide susceptibility. The spatial landslide data is one of the complex big data. To handle such as large amount of landslide data, the previous study weighted decision tree approach is improvised and Multi Ranking Decision Tree Classifier is proposed using map reduce programming model,.

Related Work

Decision trees are one of the most accepted methods for classification in diverse data mining applications (H. I. Witten & E. Frank, 2005; M. J. Berry & G. S. Linoff, 1997) and help the development of decision making(J. R. Quinlan, 1990). One of the well known decision tree algorithms is C4.5 (J. R. Quinlan, 1993; J. R. Quinlan, 1996), an expansion of basic ID3 algorithm(J. R. Quinlan,1986). However, with the growing improvement of cloud computing (M. Armbrust, A. Fox, R. Griffith, A. D. Joseph, R. Katz, A. Konwinski, G. Lee, D. Patterson, A. Rabkin, I.Stoica and M. Zaharia, 2010) as well as the big data challenge (D. Howe, M. Costanzo, P. Fey, T. Gojobori, L. Hannick, W. Hide, D.P. Hill, R. Kania, M. Schaeffer, S.S., 2008), traditional decision tree algorithms reveal numerous restrictions. First and foremost, building a decision tree can be very time consuming when the volume of dataset is extremely big, and new computing paradigm should be applied for clusters. Second, although parallel computing(V. Kumar, A. Grama, A. Gupta & G. Karypis, 1994) in clusters can be leveraged in decision tree based classification algorithms (K. W. Bowyer, L. O. Hall, T. Moore, N. Chawla & W. P. Kegelmeyer, 2000; J. Shafer, R. Agrawal & M. Mehta, 1996), the strategy of data distribution should be optimized so that required data for building one node is localized and mean while the communication cost to be minimized. Weighted classification are well-suited for many real-world binary classification problems. Weighted classification (J.L.Polo, F.Berzal, & J.C.Cubero, 2007) assigns different importance degrees to different attributes. Many different splitting criteria for attribute selection have been proposed in the literature and they all tend to provide similar results (F.Berzal, J.C.Cubero, F.Cuenca, & M.J.Martín-Bautista, 2003).

HACE theorem (Xindong Wu,Xingquan Zhu,Gong-Qing Wu, & WeiDing.S, 2014) has been presented which characterizes the features of the Big Data revolution, and proposes a Big Data processing model, from the data mining perspective. This data-driven model involves demand-driven aggregation of information sources, mining and analysis, user interest modeling, and security and privacy considerations. We analyze the challenging issues in the data-driven model and also in the Big Data revolution.

An integration of remote sensing, GIS and Data mining techniques has been used to predicting the landslide risk. The probabilistic and statistical approaches were applied for estimating the landslide susceptibility area. Landslide susceptibility map is reduced the landslide hazard and is used for land cover planning. The frequency ratio model has better than logistic regression model. Fuzzy membership functions and factor analysis were used to assess the landslide susceptibility using various factors. The spatial data were collected and processed and create a spatial database using GIS and Image processing techniques. The landslide occurrence factor was identified and processed. Each factor weight was determined and calculated the training using back-propagation. Improvised Bayesian Classification approach (Venkatesan M*, Rajawat A S, Arunkumar T, Anbarasi M, & Malarvizhi K, 2014) and decision tree approach (Venkatesan.M, Arunkumar .Thangavelu, & Prabhavathy.P, 2013) have been applied to predict the landslide susceptibility in Nilgiris district.

Multi Ranking Decision Tree Classifier

Classification is the process to predict the unknown class label using training data set. Classification approaches are categorized into Decision Tree, Back propagation Neural Network, Support Vector machine(SVM), Rule based Classification and Bayesian Classification. In the present scenario, landslide analysis study was done by using Neural Network and Bayesian but these approaches are difficult to understand and tricky to predict. In this chapter, Multi Ranking Decision Tree Classifier is proposed for landslide Risk Analysis. The performance of the proposed approach is measured with various parameters.

Decision Tree (DT) approach is used to analyze the data in the form of tree. The Tree is constructed using the top-down and recursive splitting technique. A tree structure consists of a root node, internal nodes, and leaf nodes. Ranking classification techniques give simpler models for the important classes. Ranking classification assigns different importance degrees to different landslide factor. In this chapter, rankings are assigned to the different landslide factors in order to represent the relative importance of each landslide factor. In a distributing computing environment, the large data sets are handled by an open source framework called Hadoop. It consists of MapReduce, Hadoop file distribution system (HDFS) and number of related projects Apache Hive, HBase and Zookeeper.

The Hadoop Distributed File Systems (HDFS) architecture is illustrated in Figure 1 NameNode is the master node of HDFS handling metadata, and DataNode is slave node with data storage in terms of blocks. Similarly, the Master node of Hadoop MapReduce is called JobTracker, which is in charge of managing and scheduling several tasks, and the slave node is called TaskTracker, where Map and Reduce procedures are actually performed.

MapReduce programming model is used for parallel and distributed processing of large datasets on clusters (Venkatesan M*, Rajawat A S, Arunkumar T, Anbarasi M, & Malarvizhi K, 2014). There are two basic procedures in MapReduce: Map and Reduce.

In general, the input and output are both in the form of key/value pairs. Figure 2 shows MapReduce programming model architecture. The input data is divided in to block in the size of 68MB or 128 MB. The mapper input will be supplied as key/value paris and it produces the relative output in the form of key/pairs. Partitioner

Figure 1. HDFS Architecture

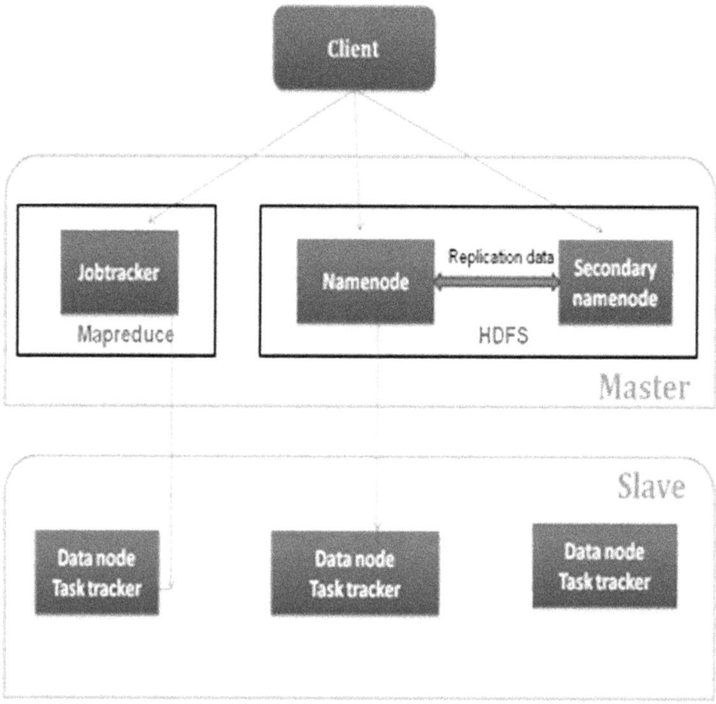

Figure 2. MapReduce Architecture

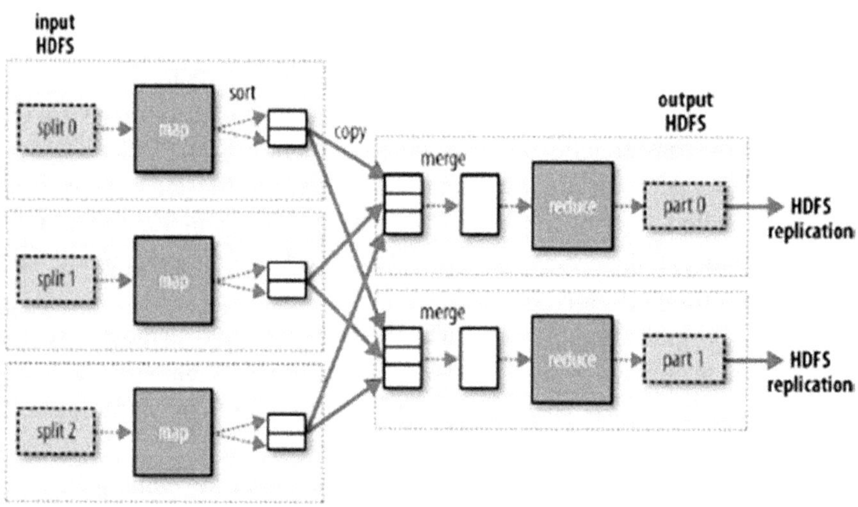

and combiner are used in between mapper and reducer to perform sorting and shuffling. The Reducer iterates through the values that are associated with specific key and produces zero or more outputs.

The dataset is relatively huge in a big data atmosphere, designing appropriate data structures for parallel programming is very much important. Three data structures such as attribute table, count table, hash table are used to build parallel decision tree classifier. Basic information of attribute a, the row identifier of instance row _ id, values of attribute values (a) and class labels of instances c are stored in attribute table . Count table computes the count of instances with specific class labels if split by attribute a. That is, two fields are included: class label c and a count cnt .The last one is hash table, which stores the link information between tree nodes node _ id and row _ id, as well as the link between parent node node _ id and its branches.

The traditional data is converted into above three data structure for MapReduce processing. The algorithm -I, procedure data conversion transforms the instance record into attribute table with attribute Aj as key, and row _ id and class label c as values. Then, REDUCE_ATTRIBUTE computes the number of instances with specific class labels if split by attribute Aj, which forms the count table.Note that hash table is set to null at the beginning of process.

Algorithm -I: Data Conversion

```
Procedure Map-Attribute (tuple_id,(A1,A2,...A3,C))
        emit(Aj,(row_id,C))
End procedure
Procedure Reduce-Attribute ((Aj,(row_id,C))
emit(Aj,(C,Cnt))
End Procedure
```

In Decision Tree Classifier, selecting best splitting attribute is important task. The algorithm - II shows that, mapper performs the computation of information and split information of Aj. The reducer computes the information gain ratio. The attribute Aj which has maximum value of GainRatio is selected as splitting attribute.

Algorithm-II: Splitting Attribute Selection

```
Procedure Reduce_Population((Aj,(C,Cnt))
        emit(Aj,all)
End Procedure
Procedure Map_Computation((Aj,(C,Cnt,all)))
        Compute Entrophy(Aj)
```

```
Compute  Info(A_j) = (Cnt/all) Entrophy(A_j)
Compute SplitInfo(A_j) = (Cnt/all) log (Cnt/all)
emit(A_j, Info(A_j), SplitInfo(A_j))
End Procedure
Procedure Reduce_Computation()
emit(A_j, GainRatio(A_j))
End Procedure
```

As shown in algorithm - III, the records are read from attribute table with key value equals to a_{best} and emit the count of class labels.

Algorithm-III: Hash Table Updation

```
Procedure Map_Update_Count((A_best, (row_id,C)))
Emit(A_best, (C,Cnt'))
End Procedure
Procedure Map_Hash((A_best, row_id))
        Compute node_id= hash(A_best )
        emit(row_id, node_id)
End Procedure
```

Algorithm –IV shows the procedure to grow the decision tree by building linkages between nodes.

Algorithm-IV: Building Tree

```
Procedure Map((A_best,row_id))
Compute node_id=hash(A_best)
If node_id is same with the old value then
        emit(row-id,node_id)
end if
add a new subnode
emit(row_id,node_id,subnode_id)
End Procedure
```

Ranking classification techniques give simpler models for the important classes. Ranking classification assigns different importance degrees to different landslide factor. In this chapter, rankings are assigned to the different landslide factors in order to represent the relative importance of each landslide factor. Weighted decision tree classification algorithm is improved as multi ranking decision tree classifier using map reduce programming model as shown in the above algorithms. The developed classifier is used to analyze the landslide risk in the ooty region of Niligiris district. The proposed classifier scalability is improved and performance is compared with the existing classification methods.

Experiment and Result Analysis

The proposed multi ranking decision tree classifier is implemented in Hdoop cluster. We have HPC cluster with 6 nodes.We let one of them as HDFS NameNode and MapReduce JobTracker (i.e., master), and the remaining nodes act as HDFS DataNode and MapReduce TaskTracker (i.e., slave).The efficiency of weighted decision tree classification algorithm is theoretically and empirically proved in our previous study. In this paper, we are

concerned with the time efficiency of parallel version of weighted decision tree classification algorithm in big data environment. This chapter focuses landslide risk analysis using big data computational techniques. The needed toposheets and required maps are collected from the geological survey of India. Many number of factors causes landslide in the hill region, but four factors are very important for landslide study such as rainfall, slope, geology, and landuse/landcover. The above said factors thematic layers are prepared from the LISS III+ PAN images using ArcGIS Tool. Ooty, Nilgris district is considered as study area. We have applied the proposed multi ranking decision tree classifier on ooty landslide data as shown in table 1.

Table 1. Sample Landslide Data

Land use	Geology	Rainfall	Slope	Zone
Agriculture	Ultrabasic rocks	135.63-150.82	11.76-19.79	Low
Agriculture	Gneiss	135.63-150.82	8.02-11.76	Low
Agriculture	Ultrabasic rocks	135.63-150.82	8.02-11.76	Very Low
Scrub Forest	Gneiss	135.63-150.82	0-8.02	Very Low
Scrub Forest	Ultrabasic rocks	135.63-150.82	0-8.02	Very Low
Scrub Forest	Gneiss	135.63-150.82	11.76-19.79	Low
Scrub Forest	Ultrabasic rocks	135.63-150.82	11.76-19.79	Very Low

The proposed multi ranking decision tree classifier is applied on ooty landslide data and the landslide risk level is analyzed and it is shown in Figure 3.

The performance of proposed classifier is compared with the weighted decision tree classifier and decision tree classifier on single node. Figure 4 illustrates the following observations.

First, the larger the dataset is, the more time consuming it is to build the normal decision tree approach. Second, the execution time of weighted decision tree classification takes more time than proposed parallel weighted decision tree classifier. The proposed MapReduce based multi ranking classifier algorithm takes less time the original decision tree as the size of dataset increases. Therefore, it is proved that the proposed multi ranking decision tree classifier outperforms the sequential version even on a single node environment.

The scalability of the proposed ranking decision tree classification is also tested in distributed parallel domain. The scalability evaluation includes two aspects: (1)

Figure 3. Landslide Risk analysis using multi ranking decision tree classifier

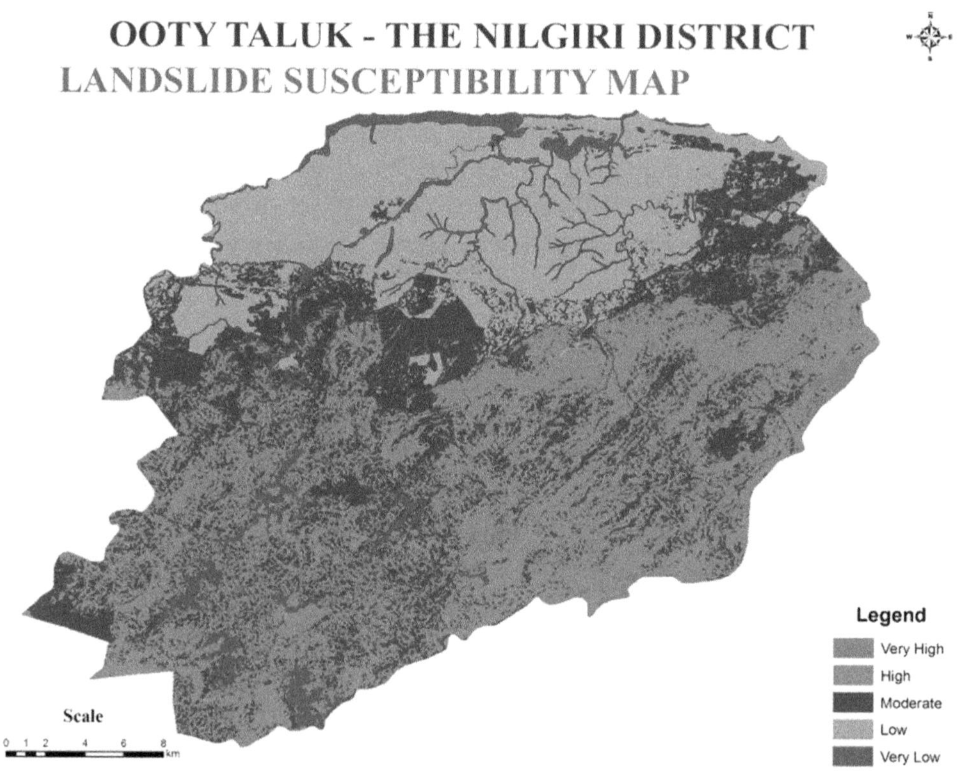

**For a more accurate representation see the electronic version.*

Figure 4. Performance of Multi Ranking Decision Tree Classifier

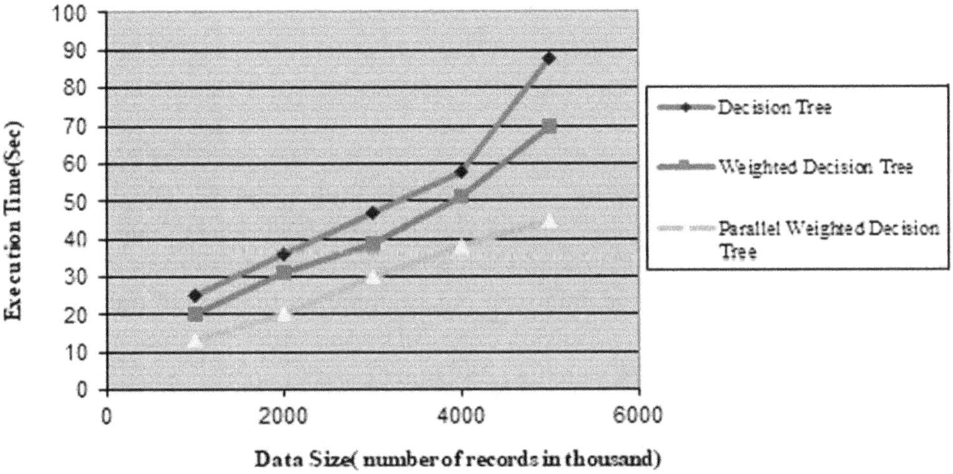

performance with different numbers of nodes, and (2) performance with different size of training datasets.

Figure 5 illustrates the execution time of our proposed weighted decision tree classification algorithm with different numbers of nodes when the number of record is 1, 2 and 3 lakhs respectively, We have observe that the overall execution time

Figure 5. Performance of Multi Ranking Decision Tree Classifier based on number of nodes

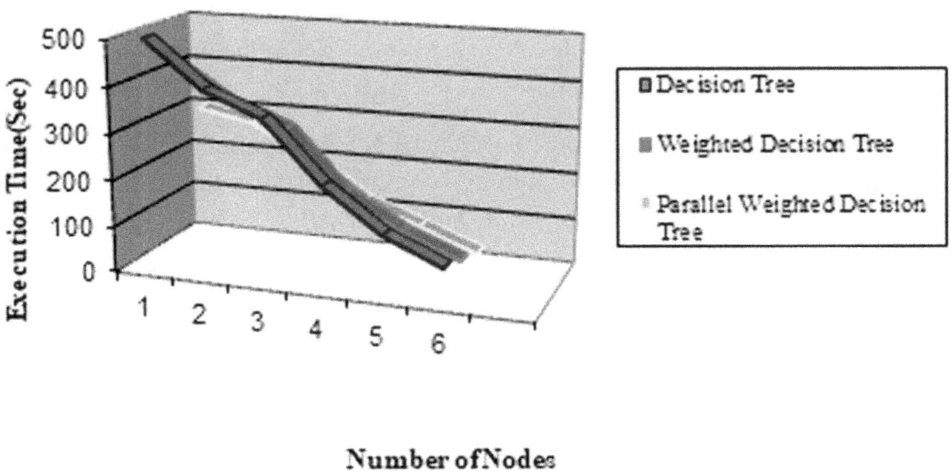

*For a more accurate representation see the electronic version.

decreases when the number of nodes increases. This indicates that the more nodes are involved for computing increases the efficiency of the algorithm.

CONCLUSION

Predicting and analyzing disaster is complex task. In this chapter, landslide risk is analyzed using multi ranking decision tree classifier approach. Disaster management domain generates huge amount of data. Traditional sequential decision tree algorithms cannot fit to handle such huge data sets. For example, as the size of training data grows, the process of building decision trees can be very time consuming. To solve the above challenges, parallel weighted decision classifier approach is proposed to improve the scalability of the model. We have compared the performance of the proposed approach with existing approach with respect to number of nodes and number of record. The empirical results shows that the proposed algorithm exhibit both time efficiency and scalability. In future works, the rainfall induced landslide risk analysis will be studied using big data computational approaches.

REFERENCES

Armbrust, Fox, Griffith, Joseph, & Katz, Konwinski, … Zaharia. (2010). A view of cloud computing. *Communications of the ACM*, *53*(4), 50–58.

Berry, M. J., & Linoff, G. S. (1997). *Data mining techniques: For marketing, sales, and customer support*. John Wiley & Sons, Inc.

Berzal, F., Cubero, J. C., Cuenca, F., & Martín-Bautista, M. J. (2003). On the quest for easy-to-understand splitting rules. *Data & Knowledge Engineering*, *44*(1), 31–48. doi:10.1016/S0169-023X(02)00062-9

Bowyer, K. W., Hall, L. O., Moore, T., Chawla, N., & Kegelmeyer, W. P. (2000), A parallel decision tree builder for mining very large visualization datasets. *IEEE International Conference on Systems Man, and Cybernetics*, 3, 1888-1893. 10.1109/ICSMC.2000.886388

Howe, D., Costanzo, M., Fey, P., Gojobori, T., Hannick, L., Hide, W., ... Rhee, S. Y. (2008). Big data: The future of biocuration. *Nature*, *455*(7209), 47–50. doi:10.1038/455047a PMID:18769432

Kumar, V., Grama, A., Gupta, A., & Karypis, G. (1994). *Introduction to parallel computing* (Vol. 110). Redwood City: Benjamin/Cummings.

Polo, J. L., Berzal, F., & Cubero, J. C. (2007). Taking class importance into account. *Lecture Notes in Computer Science*, 4413.

Quinlan. (1996). *Improved use of continuous attributes in C4.5.* arXiv preprint cs/9603103

Quinlan, J. R. (1986). Induction of decision trees. *Machine Learning*, *1*(1), 81–106. doi:10.1007/BF00116251

Quinlan, J. R. (1990). Decision trees and decision-making. *IEEE Transactions on Systems, Man, and Cybernetics*, *20*(2), 339–346. doi:10.1109/21.52545

Quinlan, J. R. (1993). *C4.5: programs for machine learning*. Morgan Kaufmann.

Shafer, J., Agrawal, R., & Mehta, M. (1996). PRINT: A scalable parallel classifier for data mining. *Proc. Int.Conf. Very Large Data Bases.*

Venkatesan, Rajawat, Arunkumar, Anbarasi, & Malarvizhi. (2014). GIS Based Data Mining Classification Approaches for Landslide Susceptibility Analysis. *International Journal of Applied Environmental Sciences*, *9*(5), 2345–2357.

Venkatesan, Arunkumar, Thangavelu, & Prabhavathy. (2013). An Improved Bayesian Classification Data mining Method for Early Warning Landslide Susceptibility Model Using GIS. *Proceedings of Seventh International Conference on Bio-Inspired Computing: Theories and Applications (BIC-TA 2012).* Springer.

Witten, H. I., & Frank, E. (2005). Data Mining: Practical machine learning tools and techniques. Morgan Kaufmann.

Wu, Zhu, & Wu, & Ding. (2014). Data Mining with Big Data. *IEEE Transactions on Knowledge and Data Engineering*, *26*(1).

Zikopoulos, P., & Eaton, C. (2011). *Understanding big data: Analytics for enterprise class hadoop and streaming data*. McGraw-Hill Osborne Media.

Chapter 3
Modified Support Vector Machine Algorithm to Reduce Misclassification and Optimizing Time Complexity

Aditya Ashvin Doshi
VIT University, India

Prabu Sevugan
VIT University, India

P. Swarnalatha
VIT University, India

ABSTRACT

A number of methodologies are available in the field of data mining, machine learning, and pattern recognition for solving classification problems. In past few years, retrieval and extraction of information from a large amount of data is growing rapidly. Classification is nothing but a stepwise process of prediction of responses using some existing data. Some of the existing prediction algorithms are support vector machine and k-nearest neighbor. But there is always some drawback of each algorithm depending upon the type of data. To reduce misclassification, a new methodology of support vector machine is introduced. Instead of having the hyperplane exactly in middle, the position of hyperplane is to be change per number of data points of class available near the hyperplane. To optimize the time consumption for computation of classification algorithm, some multi-core architecture is used to compute more than one independent module simultaneously. All this results in reduction in misclassification and faster computation of class for data point.

DOI: 10.4018/978-1-5225-3643-7.ch003

INTRODUCTION

These days, numerus organizations are using "big data", "machine learning" technologies for data analysis. These are the terms which describes that available data is so complex as well as large so that it becomes distinctly clumsy to work with existing statistical algorithms which restricts size and type of data. The existing data mining algorithm usually can be divided in to sub types like, "associate rule mining", "classification", "clustering" (A Survey on Feature Selection Techniques and Classification Algorithms for Efficient Text Classification.) Classification technique works with association of unstructured to well-structured data. Numerous amount of classification techniques is introduced in the fields of big data. As every algorithm has its own pros and cons depending upon type of data need to be classified. The performance of these techniques is generally measured in terminology of cost and cost is nothing but required computation time and misclassification.

What is Machine Learning?

Consider "machine learning" this way. As a human being, and as a client of innovation, you finish certain errands that oblige you to choose or group something. For example, when you read your inbox in the morning, you choose to stamp that "Win a Free Cruise on the off chance that You Click Here" email as spam. How might a computer know to do a similar thing? Machine learning is involved algorithms that instruct computers to perform assignments that individuals do day by day.

The primary endeavour's at counterfeit consciousness included instructing a computer by composing a run the show. On the off chance that we needed to educate a computer to make proposals considering the climate, then we may compose a decide that stated: IF the climate is shady AND the possibility of precipitation is more prominent than half, THEN recommend taking an umbrella. The issue with this approach utilized as a part of conventional master frameworks, in any case, is that we don't know how much certainty to put on the run the show.

Hence, machine learning has developed to imitate the example coordinating that human brains perform. Today, machine learning algorithms instruct computers to perceive components of a protest. In these models, for instance, a computer is demonstrated an apple and told that it is an apple. The computer then uses that data to characterize the different attributes of an apple, expanding upon new data each time. At initial, a computer may arrange an apple as round, and fabricate a model that expresses that if something is around, it's an apple. At that point later, when an orange is presented, the computer discovers that if something is around AND red, it's an apple. At that point a tomato is presented, etc. The computer should persistently alter its model considering new data and allot a prescient incentive to each model, showing the level of certainty that a question is one thing over another. For instance, yellow is a more prescient incentive for a banana than red is for an apple.

So Why is Everyone Talking about Machine Learning?

These essential algorithms for instructing a machine to finish assignments and order like a human go back a very long while. The contrast amongst now and when the models were initially designed is that the more data is sustained into the algorithms, the more precise they move toward becoming. The previous couple of decades have seen enormous versatility of information and data, taking into consideration a great deal more precise forecasts than were ever conceivable in the long history of machine learning.

New systems in the field of machine learning – that generally include consolidating pieces that as of now existed in the past – have empowered a remarkable research exertion in "Deep Neural Networks (DNN)". This has not been the consequence of a noteworthy leap forward, yet rather of substantially speedier computers and a great many analysts contributing incremental upgrades. This has empowered scientists to extend what's conceivable in machine learning, to the point that machines are beating people for troublesome yet barely characterized errands, for example, perceiving appearances or playing the round of Go.

Why is this Important?

Machine learning has a few exceptionally functional applications that drive the sort of genuine business comes about-- for example, time and cash funds – that can possibly significantly affect the eventual fate of your association. At Connections, specifically, we see colossal effect happening inside the client mind industry, whereby machine learning is permitting individuals to accomplish things more rapidly and proficiently. Through virtual right-hand arrangements, machine learning computerizes errands that would some way or another should be performed by a live operator –, for example, changing a secret key or checking a record adjust. This arranges for significant operator time that can be utilized to concentrate on the sort of client care that people perform best: high touch, confused basic leadership that is not as effectively dealt with by a machine. At Associations, we additionally enhance the procedure by wiping out the choice of whether a demand ought to be sent to a human or a machine: one of a kind versatile understanding innovation, the machine figures out how to know about its constraints, and safeguard to people when it has a low trust in giving the right arrangement.

In "Support Vector Machine", decision boundaries are defined depending upon decision plane. A plane that separate data points of different classes (Ali AlShaari,

M., 2014). The example is illustrated below in Figure 1. In this example, there are data points of two classes which are represented with two different color. In this figure one side having data points of same class and other side having data points of another class which is separated by line. New data points class can be decided depending upon that data point belong to which side of that separating line. The classification algorithm is of two types "linear classification" and "non-linear classification". "Classification" will be linear if and only if there exist a clear straight line which separates data points into distinct group, then that is nothing but linear classification (Guang-chao, W., 2008). If there is no clear straight line separating two different classes, then that is non-linear classification.

The above is a classic example of a linear classifier, i.e., a classifier that separates group of data points in to their respective category (GREEN and RED in this case) with a line. Most of the classification work is more complex structured and making optimal separation is needed and this is not that simple, i.e., correctly classify new data point based on the existing data points with known class. This situation is depicted in the illustration below. Compared to the previous schematic, a full separation of the GREEN and RED objects would require a curve (which is more complex than a line). Classification tasks based on drawing separating lines to differentiate among data points of different class member are known as hyperplane classifiers. "Support Vector Machines" are particularly suited to handle such tasks.

The representation beneath demonstrates the essential thought behind "Support Vector Machine". Here we see the original data points are mapped, i.e., adjusted, utilizing an of mathematical functions, known as kernels. The way toward reworking the articles is known as mapping (change) (Gumus, F., 2014). Take note of that in this new setting, the mapped data points are linearly divisible and, in this manner, rather than building the complex curve Figure 2 (left schematic), we should simply to locate an ideal line that can isolate the GREEN and the RED data points.

Figure 1. SVM Classifier

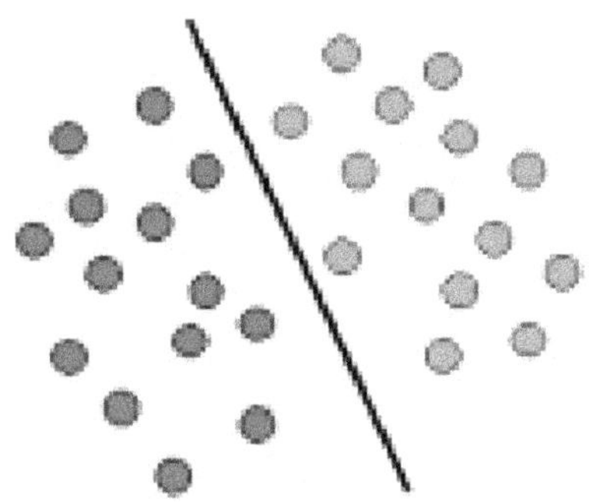

**For a more accurate representation see the electronic version.*

Figure 2. SVM Classifier

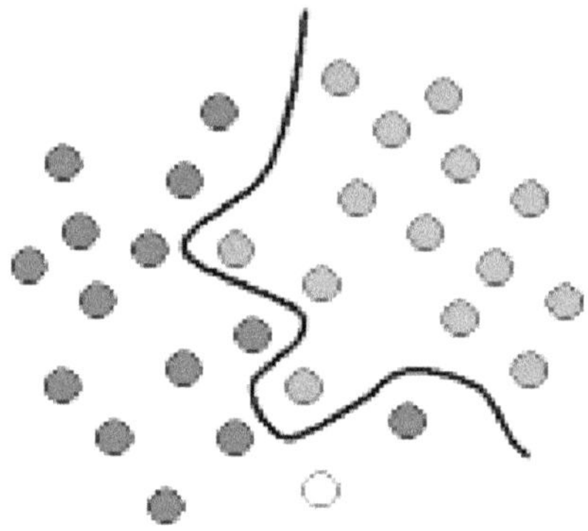

**For a more accurate representation see the electronic version.*

BACKGROUND

Classification technique works with association of unstructured to well-structured data. Numerous amount of classification techniques is introduced in the fields of big data. As every algorithm has its own pros and cons depending upon type of data need to be classified. The performance of these techniques is generally measured in terminology of cost and cost is nothing but required computation time and misclassification.

The Role of Classification in Data Mining

In view of the information gathered, information mining algorithms are utilized to either create a depiction of the information put away, or anticipate an outcome (Laskov, P., & Lippmann, R., 2010). Various types of algorithms are utilized to accomplish both assignments. Nonetheless, in the general KDD prepare, any blend of these assignments might be called upon to accomplish the coveted outcomes. The Steps required in KDD are:

1. **Description Errands:** These assignments depict the information being mined and they are:
    a. **Summarization:** To concentrate minimal patterns that portray subsets of information. The technique used to accomplish this undertaking are Association Rule algorithms.
    b. **Segmentation or Clustering:** To separate information things into subsets that are like each other. Segment based grouping algorithms are utilized to accomplish this undertaking.

   c. **Change and Deviation Detection:** To distinguish changes in successive information, (for example, protein sequencing, behavioral groupings, and so on.).
   d. **Dependency Modeling:** To build models of causality inside the information.
2. **Prediction Assignments:** To foresee some field(s) in a database in light of data in different fields.
   a. **Classification:** To foresee the in all likelihood condition of a clear cut variable (its class).
   b. **Regression:** To foresee comes about that are numeric persistent factors.

Classification

- **Model Construction:** Model development is building the model from the preparation set.
  - Each tuple/test is expected to have a place a prefined class
  - The class of a tuple/test is dictated by the class name quality
  - The preparation set of tuples/tests is utilized for model development
  - The model is spoken to as arrangement principles, choice trees or scientific formulae
- Model Usage
  - Group future or obscure items
  - Appraise exactness of the model
  - The known class of a test tuple/test is contrasted and the outcome given by the mode
  - Accurate rate = rate of the tests tuples/tests effectively arranged by the model

Existing Methods in Data Classification

With a colossal measure of information put away in databases and information distribution centres, it is progressively imperative to grow capable apparatuses for investigation of such information and mining intriguing learning from it. Information mining is a procedure of construing learning from such tremendous information. Information Mining has three noteworthy parts Clustering or Classification, Association Rules and Sequence Analysis. Information Classification is a critical stride in information mining applications.

By straightforward definition, in characterization/bunching we break down an arrangement of information and produce an arrangement of collection principles which can be utilized to group future information. For instance, one may order infections and give the side effects which portray each class or subclass. This has much in a similar manner as conventional work in measurements and machine learning. In any case, there are vital new issues which emerge due to the sheer size of the information. One of the essential issue in information mining is the Classification-lead learning which includes discovering decides that parcel given information into predefined classes. In the information mining space where a great many records and countless are included, the execution time of existing algorithms can wind up noticeably restrictive, especially in intuitive applications. In Data characterization one builds up a portrayal or model for each class in a database, in view of the components exhibit in an arrangement of class-named preparing data. There have been numerous information arrangement techniques, for example, choice tree strategies, for example, statistical techniques, neural systems, harsh sets, database-situated techniques and so forth.

Data Classification Methods

The accompanying rundown demonstrates the accessible information arrangement techniques.

- **Statistical Algorithms:** Statistical examination frameworks, for example, SAS and SPSS have been utilized by experts to identify surprising patterns and clarify patterns utilizing statistical models, for example, direct models. Such frameworks have their place and will keep on being utilized.
- **Neural Networks:** Artificial neural systems copy the pattern-discovering limit of the human cerebrum and consequently a few specialists have recommended applying Neural Network algorithms to pattern mapping. Neural systems have been connected effectively in a couple of utilizations that include grouping.
- **Genetic Algorithms:** Optimization systems that utilization procedures, for example, hereditary blend, transformation, and normal determination in an outline in view of the ideas of common development.
- **Nearest Neighbour Method:** A procedure that orders each record in a dataset considering a blend of the classes of the k record(s) most like it in a verifiable dataset. Here and there called the k-closest neighbour method.
- **Rule Enlistment**: The extraction of helpful if-then principles from information considering statistical importance.
- **Data Visualization:** The visual elucidation of complex connections in multidimensional information.

Many existing methods propose abstracting the test information before arranging it into different classes. There are a few options for doing reflection before order: An informational index can be summed up to either a negligibly summed up deliberation level, a transitional deliberation level, or a high deliberation level. Too low a deliberation level may bring about scattered classes, thick grouping trees, and trouble at compact semantic understanding; though too high a level may bring about the loss of characterization exactness.

1. Genetic Algorithm

Development has turned out to be an effective instrument in discovering great answers for troublesome issues. One can take a gander at the normal determination as an advancement strategy, which tries to create sufficient answers for specific conditions.

Disregarding the expansive number of uses of GA in various sorts of enhancement issues, there is almost no examination on utilizing this sort of way to deal with the grouping issue. And remembering the nature of the arrangements that this innovation has appeared in changed sorts of fields and issues (Beasley, Bull & Martin, 1993a, & Mitchell, 1996) it bodes well to attempt to utilize it in grouping issues.

The adaptability related with GA is one critical perspective to hold up under as a top priority. With a similar genome portrayal and just by changing the wellness work one can have an alternate algorithm. Because spatial investigation this is especially vital since one can attempt distinctive wellness works in an exploratory stage.

In the genome, every quality speaks to an information point and characterizes group enrolment. All vital development administrators can be executed with this plan. As pointed by Demiriz (1999) each of the real issue related with this portrayal plan is that it is not adaptable, then again it is by all accounts computationally effective when the quantity of information focuses is not very expansive.

2. Decision Trees

A choice tree is a characterization plan which produces a tree and an arrangement of principles, speaking to the model of various classes from a given information set. The arrangement of records accessible for creating characterization strategies is for the most part partitioned into two disjoint subsets as takes after:

a. **A preparing set:** Utilized for inferring the classifier.
b. **A test set:** Used to gauge the precision of the classifier.

The precision of the classifier is dictated by the rate of the test cases that are accurately arranged. The characteristics of the records are partitioned into two sorts as takes after:

a. **Numerical traits:** Qualities whose space is numerical.
b. **Clear cut qualities:** Traits whose area is not numerical.

There is one recognized trait called the class mark. The objective of the arrangement is to assemble a compact model that can be utilized to anticipate the class of the records whose class mark is not known.

A choice tree is a tree where the interior hub - is a test on a quality, the tree limb - is a result of the test, and the leaf hub - is a class name or class dispersion.

There are two periods of choice tree generation:

Tree Construction

- At begin, all the preparation cases are at the root
  - Partition cases considering chose qualities
  - Test traits are chosen considering a heuristic or a statistical measure

Tree pruning

- Recognize and expel branches that reflect clamour or anomalies
  - One govern is produced for every way in the tree from the root to a leaf
  - Each characteristic esteem match along a way structures a conjunction
  - The leaf hub holds the class expectation
  - Rules are for the most part less difficult to comprehend than trees

Tree Construction Principle

There are different techniques for building choice trees from a given preparing informational index. Some essential ideas required in the working of choice trees are examined beneath.

Splitting Attribute

With each hub of the choice tree, there is a related property whose qualities decide the apportioning of the informational index when the hub is extended.

Splitting Criterion

The qualifying condition on the part property for informational collection part at a hub is known as the part foundation at that hub. For a numeric characteristic, the paradigm can be a condition or an imbalance. For a clear-cut trait, it is a participation condition on a subset of qualities.

3. Decision Tree Construction Algorithms

Various algorithms for instigating choice trees have been proposed throughout the years. They vary among themselves in the techniques utilized for choosing part properties and part conditions. These algorithms can be arranged into two sorts. The principal kind of algorithms is the established algorithms which handle just memory occupant information. The second class can deal with the proficiency and adaptability issues. These algorithms evacuate the memory limitations and are quick and versatile.

4. Naïve K-Implies algorithm

A standout amongst the most prominent heuristics for taking care of the k-means issue depends on a basic iterative plan for finding a locally ideal arrangement. This algorithm is regularly called the k-implies algorithm. There are various variations to this algorithm, so to illuminate which rendition we are utilizing, we will allude to it as the credulous k-implies algorithm as it is substantially less complex contrasted with alternate algorithms portrayed here.

The guileless k-implies algorithm segments the dataset into "k" subsets with the end goal that all records, starting now and into the foreseeable future alluded to as focuses, in each subset "have a place" to a similar focus. Likewise, the focuses in each subset are nearer to that inside than to some other focus. The parcelling of the space can be contrasted with that of Veronesi dividing aside from that in Veronesi apportioning one segments the space in view of separation and here we segment the focuses considering separation. The algorithm monitors the centroids of the subsets, and continues in straightforward emphases. The underlying dividing is haphazardly created, that is, we arbitrarily instate the centroids to a few focuses in the locale of the space. In every cycle step, another arrangement of centroids is created utilizing the current arrangement of centroids taking after two extremely straightforward strides. Give us a chance to mean the arrangement of centroids after the ith emphasis by

The accompanying operations are performed in the means:

- Partition the focuses in view of the centroids C (i), that is, discover the centroids to which each of the focuses in the dataset has a place. The focuses are apportioned in view of the Euclidean separation from the centroids.
- Set another centroid c(i+1) ∈ C (i+1) to be the mean of the considerable number of focuses that are nearest to c(i) ∈ C (i) The new area of the centroid in a specific segment is alluded to as the new area of the old centroid.

The algorithm is said to have joined while re-computing the allotments does not bring about an adjustment in the apportioning. In the wording that we are utilizing, the algorithm has merged totally when C(i) and C(i – 1) are indistinguishable. For designs where no point is equidistant to more than one focus, the above joining condition can simply be come to. This meeting property alongside its straightforwardness adds to the engaging quality of the k-implies algorithm.

The naive k-means needs to play out countless "neighbour" inquiries for the focuses in the dataset. If the information is "d" dimensional and there are "N" focuses in the dataset, the cost of a solitary emphasis is O(kdN). As one would need to run a few cycles, it is for the most part not achievable to run the innocent k-means algorithm for expansive number of focuses.

Here and there the union of the centroids (i.e. C(i) and C(i+1) being indistinguishable) takes a few cycles. Likewise, in the last a few emphases, the centroids move practically nothing. As running the costly cycles such many more circumstances won't not be proficient, we require a measure of joining of the centroids with the goal that we stop the emphases when the merging criteria is met. Twisting is the most broadly acknowledged measure.

Bunching mistake measures a similar rule and is occasionally utilized rather than bending. Truth be told k-means algorithm is intended to improve twisting. Putting the bunch focus at the mean of the considerable number of focuses limits the bending for the focuses in the group. Likewise, when another group focus is more like a point than its present bunch focus, moving the group from its present bunch to the next can lessen the bending further. The over two stages are correctly the means done by the k-means group. Along these lines k-means decreases twisting in each progression locally. The k-Means algorithm ends at an answer that is locally ideal for the twisting capacity. Thus, a characteristic decision as a meeting standard is contortion. Among different measures of merging utilized by different analysts, we can quantify the total of Euclidean separation of the new centroids from the old centroids. In this proposition, we generally utilize grouping blunder/twisting as the union rule for all variations of k-means algorithm.

5. The Greedy K-Means Algorithm

The nearby union properties of k-means have been enhanced in this algorithm. Likewise, it doesn't require the underlying arrangement of centroids to be chosen. The thought is that the worldwide minima can be come to through a progression of neighbourhood ventures in view of the worldwide bunching with one group less.

Assumption: The presumption utilized as a part of the algorithm is that the worldwide optima can be come to by running k-means with the (k-1) clusters being put at the ideal positions for the (k-1) grouping issue and the kth bunch being set at a fitting position that is yet to be found.

Give us a chance to accept that the issue is to discover K clusters and K' $\leq$ K. We use the above suspicion, the worldwide optima for k = K' clusters is registered as a progression of neighbourhood quests. Expecting that we have tackled the k-means bunching issue for K' – 1 clusters, we need to put another group at a fitting area. To find the fitting addition area, which is not known, we run k-means algorithm until joining with each of the focuses in the whole arrangement of the focuses in the dataset being included as the applicant new bunch, each one in turn, to the K' – 1 clusters. The focalized K clusters that have the base mutilation after the joining of k-means in the above neighbourhood quests are the clusters of the worldwide k-means. We realize that for k = 1, the ideal grouping arrangement is the mean of the considerable number of focuses in the dataset. Utilizing the above technique, we can figure the ideal positions for the k = 2, 3, 4, ... K, clusters. Subsequently the procedure includes figuring the ideal k-means communities for each of the K = 1, 2, 3… K clusters. The algorithm is completely deterministic.

Although the engaging quality of the worldwide k-means lies in it finding the worldwide arrangement, the technique includes a substantial cost. K-means is run N times, where N is the quantity of focuses in the dataset, for each group to be embedded. The unpredictability can be decreased significantly by not running the K-means with the new group being embedded at each of the dataset focuses however by finding another arrangement of focuses that could go about as a proper set for inclusion area of the new bunch.

The variation of the kd-tree parts the focuses in a hub utilizing the plane that goes through the mean of the focuses in the hub and is opposite to the essential segment of the focuses in the hub. A hub is not part if it has not exactly a pre-indicated number of focuses or an upper bound to the quantity of leaf hubs is come to. The thought is that regardless of the possibility that the kd-tree were not utilized for closest neighbour questions, only the development of the kd-tree in light of this system would give a decent preparatory bunching of the information. We can accordingly utilize the kd-tree hubs focuses as the applicant/introductory inclusion positions for the new clusters. The time multifaceted nature of the algorithm can likewise be enhanced by adopting an eager strategy. In this approach, running k-means for every conceivable inclusion position is stayed away from. Rather diminishment in the contortion when the new group is included is considered without running k-means. The point that gives the most extreme reduction in the mutilation when included as a group focus is taken to be the new addition position.

K-means is keep running until union on the new rundown of clusters with this additional point as the new group. The presumption is that the point that gives the most extreme abatement in twisting is likewise the point for which the focalized clusters would have the minimum mutilation. These outcomes in a considerable change in the running time of the algorithm, as it is pointless to run k-means for all the conceivable inclusion positions. Be that as it may, the arrangement may not be all inclusive ideal but rather an inexact worldwide arrangement.

6. Self-Organizing Map

The SOM can be described as:

an unsupervised system that tries to take in a ceaseless topological mapping of an arrangement of sources of info onto an arrangement of yields such that the yields obtain an indistinguishable topological request from the contributions, by means of self-association in light of information cases (Openshaw & Wymer 1994).

Neurons are ordinarily sorted out in a 2D matrix, and the SOM tries to discover clusters to such an extent that any two clusters that are near each other in the lattice space have codebook vectors that are near each other in the information space.

In the self-association prepare the information vectors are introduced to the system, and the group unit whose weight vector is nearest (as a rule as far as Euclidean separation) is picked as the victor. The following stride is to refresh the estimation of the triumphant unit and neighbouring units, this will inexact the estimations of the units to the one of the information vector. This be a movement of the units toward the info vector, the extent of this development relies on upon the learning rate, which diminishes along the procedure to get joining. Remembering that Vector Quantization (VQ) is basically the same as the k-means algorithm, and that the VQ is an exceptional instance of the SOM, in which the area size is zero, one can state that there is a nearby connection amongst SOM and k-means. Openshaw and Wymer (1994) go further and say that the fundamental SOM algorithm

... is basically the same as a K means classifier; with a couple of contrasts because of neighbouring preparing which may well be viewed as a type of recreated tempering and it might give better outcomes and dodge some nearby optima.

Machine learning techniques can also be divided as supervised leaning and unsupervised leaning.

Supervised Learning

Supervised learning is an algorithm in which both the information sources and yields can be seen. Considering this preparation information, the algorithm needs to sum up with the end goal that it can accurately react to every conceivable info (Kulesza, A., 2012). This algorithm is relied upon to create amend yield for sources of info that weren't experienced amid preparing. In supervised learning, what must be discovered is determined for every illustration. Supervised arrangement happens when a coach gives the characterization to every illustration. Supervised learning of activities happens when the operator is given prompt input about the estimation of each activity (Kulesza, A., 2012). To tackle a give issue utilizing administered learning algorithm one must take after some specific strides:

1. Determine the kind of training illustrations.
2. Gather a training set.
3. Determine the info include portrayal of educated capacity.
4. Determine the structure of learning capacity and relating learning algorithm.
5. Complete the plan and run the learning algorithm on the assemble set of information.
6. Evaluate the exactness of the educated capacity likewise the execution of the learning capacity ought to be measured and after that the execution ought to be again measured on the set which is not the same as the preparation set.

Figure 3. Supervised Learning

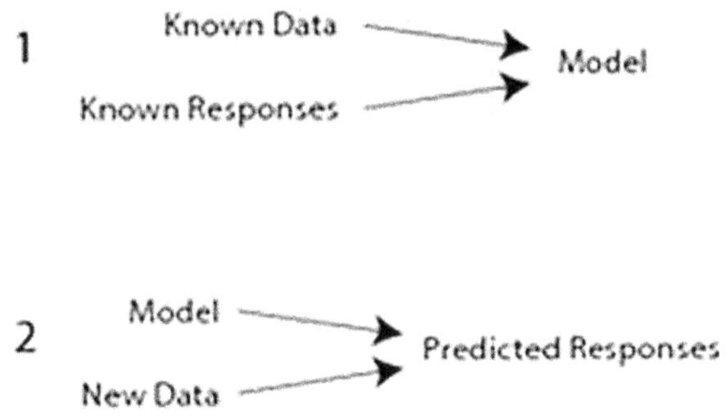

Unsupervised Learning

Unsupervised learning considers how frameworks can figure out how to speak to specific information patterns in a way that mirrors the measurable structure of the general gathering of info patterns (Laskov, P., & Lippmann, R., 2010). By appear differently in relation to "SUPERVISED LEARNING or REINFORCEMENT LEARNING", there are no express target yields or natural assessments related with each info; rather the unsupervised learner conveys to hold up under earlier predispositions concerning what parts of the structure of the information ought to be caught in the yield (Laskov, P., & Lippmann, R., 2010).

Unsupervised learning is essential since it is probably going to be a great deal more typical in the mind than supervised learning. For example, there are around photoreceptors in each eye whose exercises are always showing signs of change with the visual world and which give all the data that is accessible to demonstrate what protests there are on the planet, how they are displayed, what the lighting conditions are, and so forth. Formative and grown-up versatility are basic in creature vision (see VISION AND LEARNING) – auxiliary and physiological properties of neurotransmitters in the neocortex are known to be significantly impacted by the patterns of movement in tangible neurons that happen. Be that as it may, basically none of the data about the substance of scenes is accessible amid learning. This makes unsupervised strategies fundamental, and, similarly, permits them to be utilized as computational models for synaptic adjustment.

The main things that unsupervised learning techniques need to work with are the watched input patterns, which are regularly thought to be free specimens from a basic obscure likelihood conveyance, and some express or certain from the earlier data in the matter of what is essential. One key idea is that info, for example, the picture of a scene, has distal autonomous causes, for example, objects at given areas lit up by specific lighting (Laskov, P., & Lippmann, R. (2010). Since it is on those free causes that we regularly should act, the best portrayal for an information is in their terms. Two classes of technique have been recommended for unsupervised learning. Thickness estimation methods expressly fabricate factual models, (for example, "BAYESIAN NETWORKS") of how hidden causes could make the information. Include extraction systems attempt to separate measurable regularities (or now and again inconsistencies) straightforwardly from the information sources (Laskov, P., & Lippmann, R., 2010).

Unsupervised learning as a rule has a long and recognized history. Some early impacts were "Horace Barlow" (see Barlow, 1992), who looked for methods for portraying neural codes, "Donald MacKay" (1956), who embraced a robotic theoretic approach, and "David Marr" (1970), who made an early unsupervised learning propose about the objective of learning in his model of the neocortex. The Hebb administer (Hebb, 1949), which joins measurable techniques to neurophysiological trials on versatility, has likewise thrown a long shadow. "Geoffrey Hinton" and "Terrence Sejnowski" in designing a model of learning called the Boltzmann machine (1986), imported a considerable lot of the ideas from insights that now command the thickness estimation techniques (Grenander, 1976-1981). Highlight extraction strategies have been less broadly investigated.

Bunching gives an advantageous case. Consider the case in which the sources of info are the photoreceptor exercises made by different pictures of an apple or an orange. In the space of all conceivable exercises, these specific sources of info shape two bunches, with numerous less degrees of variety than, yellower measurement. One regular errand for unsupervised learning is to discover and describe these different, low dimensional bunches.

The bigger class of unsupervised learning techniques comprises of most extreme probability (ML) thickness estimation strategies. These depend on building parameterised models (with parameters) of the likelihood dispersion, where the types of the models (and potentially earlier appropriations over the parameters) are compelled by from the earlier data as the representational objectives.

The littler class of unsupervised learning techniques looks to find how to speak to the contributions by characterizing some quality that great 6 features have, and after that hunting down those elements in the sources of info. For example, consider the case that the yield is a direct projection of the info onto a weight vector. As far as possible hypothesis infers that most such straight projections will have Gaussian insights. Consequently, on the off chance that one can discover weights with the end goal that the projection has an exceptionally non-Gaussian (for example, multi-modular) conveyance, then the yield is probably going to mirror some fascinating part of the info. This is the instinct behind a factual technique called projection interest. It has been demonstrated that projection interest can be actualized utilizing an adjusted type of Hebbian learning (Intrator & Cooper, 1992) (Laskov, P., & Lippmann, R., 2010). Orchestrating that distinctive yields ought to speak to various parts of the info ends up being shockingly precarious.

Projection interest can likewise execute a type of grouping in the case. Consider anticipating the photoreceptor exercises onto the line joining the focuses of the bunches. The circulation of all exercises will be bimodal – one mode for each bunch – and thusly exceptionally non-Gaussian. Take note of that this single projection does not describe well the nature or state of the groups.

Another case of a heuristic fundamental great elements is that causes are regularly to some degree worldwide. For example, consider the visual contribution from a protest saw top to bottom. Diverse parts of the protest may share few components, except for that they are at a similar profundity, i.e. one part of the divergence in the data from the two eyes at the different areas is comparative. This is the worldwide hidden component. By boosting the shared data amongst yields and that are figured on the premise of the different information, one can discover this difference. This strategy was designed by Becker and Hinton (1992) and is called IMAX.

Some of the Supervised Learning Methods are:

1. Support Vector Machine

Title "Support Vector Machine" (SVM) is a standout amongst the best classification algorithms in the data mining area, yet it's long preparing time constrains its utilization. The point of SVM is to discover ideal separating hyper plane with maximum margin between the two classes, which offer the good generalization capacity for future data. Their storage and computation necessities increment quickly with the quantity of preparing vectors (Ali AlShaari, M., 2014). SVM utilizes statistical learning hypothesis to boost generalization property of produced classifier.

Assume some given information indicates each have a place one of two classes, and the objective is to choose which class another information point will be in. In "Support vector machines", data point is a P-dimensional vector, and we need to know whether we can separate such data points with (P-1)- dimensional hyperplane (Ali AlShaari, M., 2014). This is known as a linear classifier. There are numerous hyperplanes that may classify the information. The best choice of hyperplane is such line with highest margin from the different data points. So, we pick the hyperplane so that the separation from it to the closest data point on each side is maximum. Such hyperplane is known as hyperplane with maximum margin.

2. K Nearest Neighbour

"K Nearest Neighbour (KNN)" is a standout amongst the most generally utilized classification algorithm in data mining, which is dependent of learning by relationship, that is by contrasting a given test tuple and existing tuples which resemble it (Dayan, P). K-NN classification decides the decision boundary locally that was produced from the need to perform segregate investigation when dependable parametric assessments of probabilistic densities are obscure or hard to decide.

The existing data cases are vectors in a multidimensional element space, each with a class mark. The training period of the algorithm comprises just of putting away the component vectors and class names of the existing data points.

In the classification stage, k is a constant which is determined by user depending upon type of data, and a non-labeled vector is grouped by appointing the name which is most regular among the k preparing tests closest to that required data point.

An ordinarily utilized metric for continuous data is "Euclidean distance". For discrete data, for example, for content classification, another metric can be utilized,

Figure 4. KNN Classifier

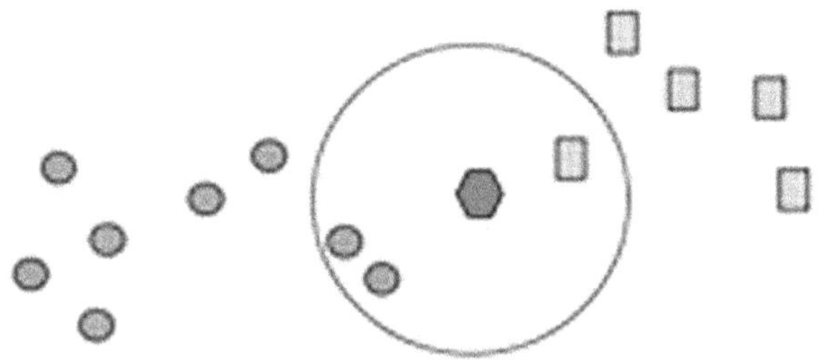

for example, the overlapping metric (or "Hamming distance") (Dubey, V., 2016). With regards to quality expression microarray data, for instance, k-NN has likewise been utilized with connection coefficients, for example, Pearson and Spearman. Frequently, the classification precision of k-NN can be enhanced fundamentally if the distance metric is found out with algorithms.

MAIN FOCUS OF THE CHAPTER

The "Support Vector Machine" (SVM) is a standout amongst the best classification algorithms in the data mining area, yet it's long preparing time constrains its utilization. The point of SVM is to discover ideal separating hyper plane with maximum margin between the two classes, which offer the good generalization capacity for future data. Their storage and computation necessities increment quickly with the quantity of preparing vectors. SVM utilizes statistical learning hypothesis to boost generalization property of produced classifier. The "K Nearest Neighbor (KNN)" is a standout amongst the most generally utilized classification algorithm in data mining, which is dependent of learning by relationship, that is by contrasting a given test tuple and existing tuples which resemble it. K-NN classification decides the decision boundary locally that was produced from the need to perform segregate investigation when dependable parametric assessments of probabilistic densities are obscure or hard to decide.

As in Figure 5 and Figure 6, a line is separated which called as hyper plane which separates the two different classes clearly. In Figure 5 the distance between the hyper plane and margin that is z1 and z2 should be equal. Hyper plane can be found by kernel function. That is nothing but it finds the maximum distance between the point's plots in vector space. That exactly middle will be the hyper plane. There is high probability that the points near to margin and hyper plan to be get misclassified. In Figure 5 the point represented in hexagon is to be classified.

Figure 5. Support vector machine

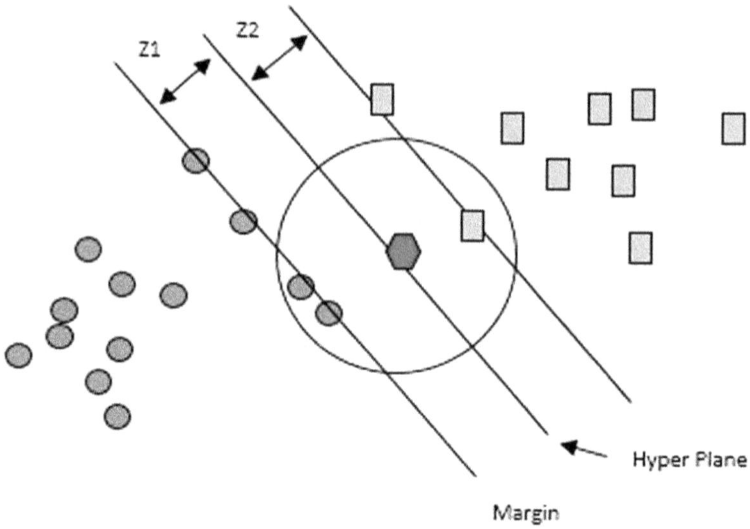

Figure 6. Proposed Support vector machine

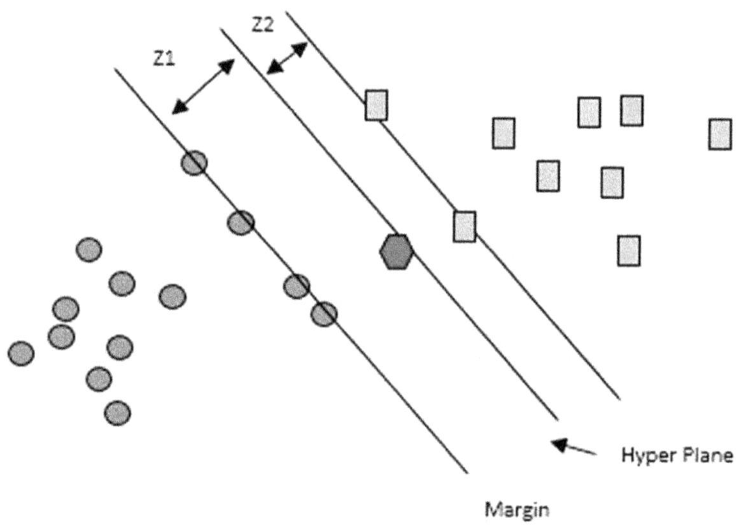

So, that point may get misclassified as it lies near to hyper plane. In modified that is in the proposed algorithm misclassification can be reduced by combining another technique to classify points near to the margin and hyperplane. That is hyper plane not necessarily in exactly middle. Hyper plane can be finding with logic for that we need get the ratio of count of data points near to both margins. Depending upon ratio hyperplane distance from both margins can be decided. Like in "k-nearest neighbor" algorithm we find the class of new data point with comparing distance of new data point with other k data points and from that k data points which class have more number of points that class will be having new data point. So, the main disadvantage of k-NN is we need to find distance of other points with respect to new data point. But in proposed methodology we only need to find ratio only once and then finalize the hyper plane.

New proposed methodology reduce misclassification without increasing the actual execution time. The independent modules of the algorithm can be executed parallel so that execution time can be reduced dramatically. This new proposed methodology will analyze the existing data points which are near to the boundary of the data points of that data class. Depending upon analysis the ratio is to be considered. Data points of which class are more near to the boundary area, then the hyper line shifted toward opposite so that misclassification can be reduced. As per existing "Support Vector Machine" in (Figure 5) the new data point represented by red colored hexagon belongs to different class of rectangles. But same data point with "K-Nearest Neighbor" belongs to circular class. But with new methodology this misclassification can be reduced as data points near to boundary area are classified using two most popular classification algorithms indirectly without increasing actual execution time.

SOLUTIONS AND RECOMMENDATIONS

Modified "support vector machine" reduces the miss classification without increasing the actual execution time. As reference to the Figure 7 and Figure 8 mentioned below there is difference in the position of the hyper plane which is separating the different class of data as represented with different colors. As there are two colored data objects are there that is red and blue which is separated by hyper plane. In Figure 7 that is existing SVM classification algorithm's scatterplot the margins are at equidistance from the margin. But inverse to the existing classification algorithm as in Figure 8 the margins are not at exactly same distance from the hyper plane. That distance will vary depending upon the number data points nearer to the hyperplane. If as in Figure 8 data point with border are considered for deciding the hyper plane and margin. Modified "support vector machine" reduces the miss classification without increasing the actual execution time. As reference to the Figure 7 and Figure 8 mentioned below there is difference in the position of the hyper plane which is separating the different class of data as represented with different colors. As there are two colored data objects are there that is red and blue which is separated by hyper

Figure 7. SVM Classifier Scatterplot

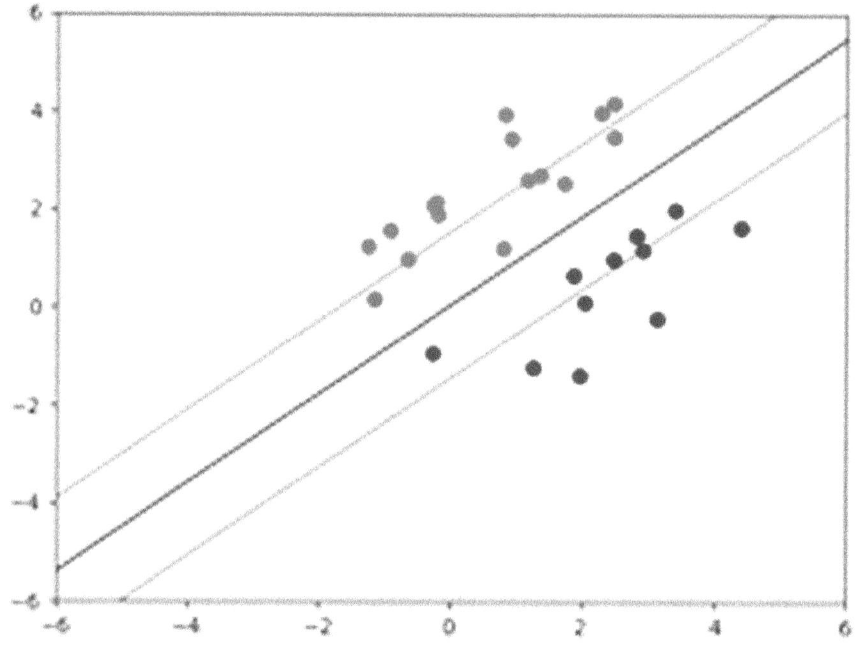

Figure 8. Modified SVM Scatterplot

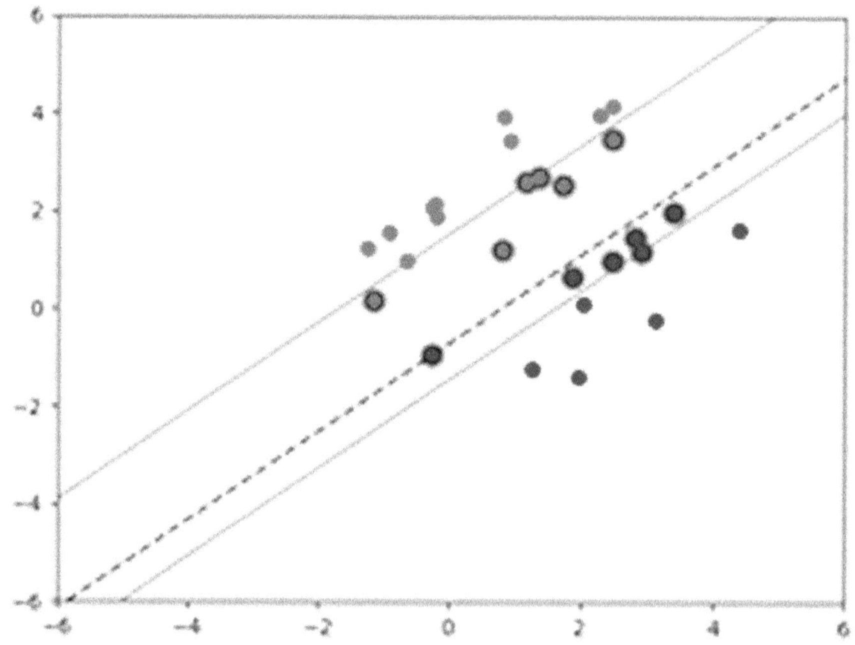

plane. In Figure 7 that is existing SVM classification algorithm's scatterplot the margins are at equidistance from the margin. But inverse to the existing classification algorithm as in Figure 8 the margins are not at exactly same distance from the hyper plane. That distance will vary depending upon the number data points nearer to the hyperplane. If as in Figure 8 data point with border are considered for deciding the hyper plane and margin.

FUTURE RESEARCH DIRECTIONS

This modified "Support vector machine" algorithm reduces the miss classification but not remove completely. In future, more classification algorithms can be added for more validation of data points. Like data points which lies near to the hyper plan have more chances to get miss classified similarly data points which lies near to boundaries are also can be easily get miss classified.

CONCLUSION

This new proposed methodology reduce misclassification as it adjusts the hyper plane depending upon data points around the border of the that class. This new algorithm is nothing but the combination of two algorithms "Support Vector Machine", "K-Nearest Neighbour". As there are high chances of misclassification of data points which lies near to the boundary of that class. So, to reduce the misclassification double verification of these sensitive area's data point is to be done through this new algorithm. As there is parallel computation of independent module is to be done, so execution time can be dramatically decreased. This new proposed algorithm works without increasing actual execution time but it reduces the misclassification.

REFERENCES

A Survey on Feature Selection Techniques and Classification Algorithms for Efficient Text Classification. (2016). *International Journal of Science and Research.*

Ali AlShaari, M. (2014). Text Documents Classification Using Word Intersections. *International Journal of Engineering and Technology*, *6*(2), 119–122. doi:10.7763/IJET.2014.V6.678

Dayan, P. (n.d.). Unsupervised Learning. Appeared. In R. A. Wilson & F. Keil (Eds.), *The MIT Encyclopedia Of The Cognitive Sciences*. MIT.

Dubey, V. (2016). *Hybrid classification model of correlation-based feature selection and support vector machine*. IEEE. doi:10.1109/ICCTAC.2016.7567338

Gayathri, K. (2013). *Data pre-processing with the KNN for classification using the SVM*. IEEE.

Guang-chao, W. (2008). Support Vector Machine Classifier Based on Fuzzy Partition and Neighborhood Pairs. *Jisuanji Yingyong*.

Gumus, F. (2014). *Online Naive Bayes classification for network intrusion detection*. IEEE. doi:10.1109/ASONAM.2014.6921657

Kulesza, A. (2012). Determinantal Point Processes for Machine Learning. Foundations And Trends® In. *Machine Learning*, *5*(2-3), 123–286. doi:10.1561/2200000044

Kulesza, A. (2012). Determinantal Point Processes for Machine Learning. Foundations And Trends® In. *Machine Learning*, *5*(2-3), 123–286. doi:10.1561/2200000044

Laskov, P., & Lippmann, R. (2010). Machine learning in adversarial environments. *Machine Learning*, *81*(2), 115–119. doi:10.100710994-010-5207-6

Talwar, A., & Kumar, Y. (2013). Article. *International Journal of Engineering and Computer Science, 2*(12).

Valsala, S., Ann George, J., & Parvathy, P. (2011). A Study of Clustering and Classification Algorithms Used in Datamining. *International Journal of Computer Science and Network Security, 11*(10).

Wang, L. (2012). *Improved KNN classification algorithms research in text categorization*. IEEE. doi:10.1109/CECNet.2012.6201850

Wang, S. (2013). *Support Vector Machines Classification for High-Dimensional Dataset*. IEEE.

Wu, G. (2008). Support vector machine classifier based on fuzzy partition and neighborhood pairs. *Jisuanji Yingyong*, *28*(1), 131–133. doi:10.3724/SP.J.1087.2008.00131

Chapter 4
An Analysis of Usage-Induced Big Data

Sameera K.
VIT University, India

P. Swarnalatha
VIT University, India

ABSTRACT

With the predominance of administration registering and distributed computing, an ever-increasing number of administrations are developing on the internet, producing tremendous volume of information. The mind-boggling administration-created information turn out to be too extensive and complex to be successfully prepared by customary methodologies. The most effective method to store, oversee, and make values from the administration-situated enormous information turn into a vital research issue. With the inexorably huge measure of information, a solitary framework that gives normal usefulness to overseeing and dissecting diverse sorts of administration-produced enormous information is critically required. To address this test, this chapter gives a review of administration-produced huge information and big data-as-a-service. Initially, three sorts of administration-produced huge information are abused to upgrade framework execution. At that point, big data-as-a-service, including big data infrastructure-as-a-Service, big data platform-as-a-service, and big data analytics software-as-a-service, is utilized to give regular huge information-related administrations (e.g., getting to benefit-produced huge information and information investigation results) to clients to improve effectiveness and lessen cost.

DOI: 10.4018/978-1-5225-3643-7.ch004

INTRODUCTION

After entering the 21st century, the worldwide financial structure is exchanging from "mechanical economy" to "administration economy". As per the insights of the World Bank, the yield of present day benefit industry takes more than 60 percent of the world yield, while the rate in created nations surpasses 70%. The opposition in the region of current benefit industry is turning into a point of convergence of the world's economy advancement. Benefit registering, which gives adaptable registering designs to bolster present day benefit industry, has developed as a promising exploration range. With the commonness of distributed computing, increasingly present day administrations are conveyed in cloud foundations to give rich functionalities. The quantity of administrations and administration clients are expanding quickly. There has been gigantic blast in information era by these administrations with the predominance of versatile gadgets, client informal organizations, and substantial scale benefit situated frameworks. The staggering administration produced information turn out to be as well expansive and complex to be viably handled by conventional approaches.

In data innovation, huge information has developed as a broadly perceived pattern, pulling in considerations from government, industry and the scholarly world. Enormous information are high volume, high speed, and additionally high assortment data resources that require new types of preparing to empower upgraded basic leadership, understanding disclosure and process enhancement (M. A. Beyer & D. Laney). Specified by the Compliance, Governance and Oversight Council (CGOC, an association concentrated on Information Governance), data volume copies each 18-24 months for most associations and 90% of the information on the planet has been made over the most recent two years (D. Austin). In March 2012, the Obama organization reported the huge information innovative work activity, which investigated how enormous information could be utilized to address imperative issues confronting the legislature. The driving IT organizations, for example, SAG, Oracle, IBM, Microsoft, SAP and HP, have spent more than $15 billion on purchasing programming firms represent considerable authority in information administration and examination. This industry all alone is worth more than $100 billion and developing at right around 10% a year, which is generally twice as quick as the product business in general ("A special report on managing information: Data, data everywhere,"). Step by step instructions to proficiently and viably make values from the huge information turn into an imperative investigate issue.

The developing extensive scale benefit situated frameworks frequently include countless with complex structures. The enormous information created from these frameworks are ordinarily heterogeneous, of various information sorts, and very powerful. Due to the quick increment of framework size and the related huge volume of administration produced information, making an incentive in the nearness of huge framework and information turns into an unavoidable test. Cases of administration created enormous information incorporate follow logs, Quality-of-Service (QoS) data, benefit summon relationship, and so forth. Like different sorts of huge information, the usage-created huge information activities traverse four remarkable measurements ("What is big data? ł bringing big data to the enterprise,"): (1) volume: these days' huge scale frameworks are flooded with constantly developing information, effortlessly gathering terabytes or even petabytes of data; (2) speed: time-touchy procedures, such as bottleneck identification and administration QoS forecast, could be accomplished as information stream into the framework; (3) assortment: organized furthermore, unstructured information are created in different information sorts, making it conceivable to investigate new experiences while breaking down these information together; and (4) veracity: identifying and remedying uproarious and conflicting information are essential to lead trustable investigation. Building up trust in enormous information displays an immense test as the assortment and number of sources develops. These four one of a kind attributes of administration created huge information give extraordinary test for information administration and examination.

To satisfy the capability of administration produced huge information, creating excellent innovations to viably handle huge amounts of information inside adequate handling time is a basic undertaking. In addition, simple access of the huge information and the enormous information examination results are imperative. Huge Data-as-a-Service embodies different enormous information stockpiling, administration, and investigation systems into administrations and gives basic enormous information related administrations to clients by means of programmable APIs, which extraordinarily upgrades proficiency, lessen cost and empowers consistent reconciliation. To give enormous information foundation, huge information stage, and huge information investigation programming projects as administrations, there are a considerable measure of examine examinations should be finished. This paper gives a diagram of administration produced huge information and Big Data-as-a-Benefit. Initial, three sorts of administration produced enormous information (benefit follow logs, benefit QoS data, and administration relationship) are misused to improve framework execution. At that point, Big Data-as-a-Service (BDaaS) is examined to give inviting APIs to clients to get to the administration created enormous information and information investigation comes about.

Whatever is left of this paper is sorted out as takes after: Section 2 gives a diagram of this paper; Section 3 misuses benefit produced enormous information; Section 4 explores Big Data-as-a-Service; Section 5 investigates business parts of usage-created huge information and Big Data-as-a-Service and Section 6 closes the paper.

ANALYSIS

Figure 1 gives an outline of the administration created enormous information and Big Data as-a-Service. As appeared in the figure, on the one hand, we will present some run of the mill applications which misuse three sorts of administration produced huge information separately for framework execution upgrade. In the first place, log representation what's more, execution program conclusion are examined by means of mining administration ask for follow logs. Second, QoS-mindful blame resilience and administration QoS forecast are concentrated in light of the benefit QoS data. At long last, noteworthy administration recognizable proof also, benefit relocation are accomplished by examining administration relationship. These solid applications will reveal some insight on the issue of enormous information investigation by mining the usage-created huge information.

Then again, to give easy to understand access to the administration produced huge information and different huge information logical comes about, Big Data-as-a-Service will likewise be presented as a fundamental structure to store, oversee and make an incentive from the huge information. Figure 1 shows the structure of Big Data-as-a-Service, which includes three layers, i.e., Big Data Infrastructure-as-a-Benefit, Big Data Platform-as-a-Service, and Big Data Analytics Programming as-a-Service. By means of standard and programmable APIs, Big Data-as-a-Service

Figure 1. Analysis of Usage-induced Big Data and Big Data-as-a-Service

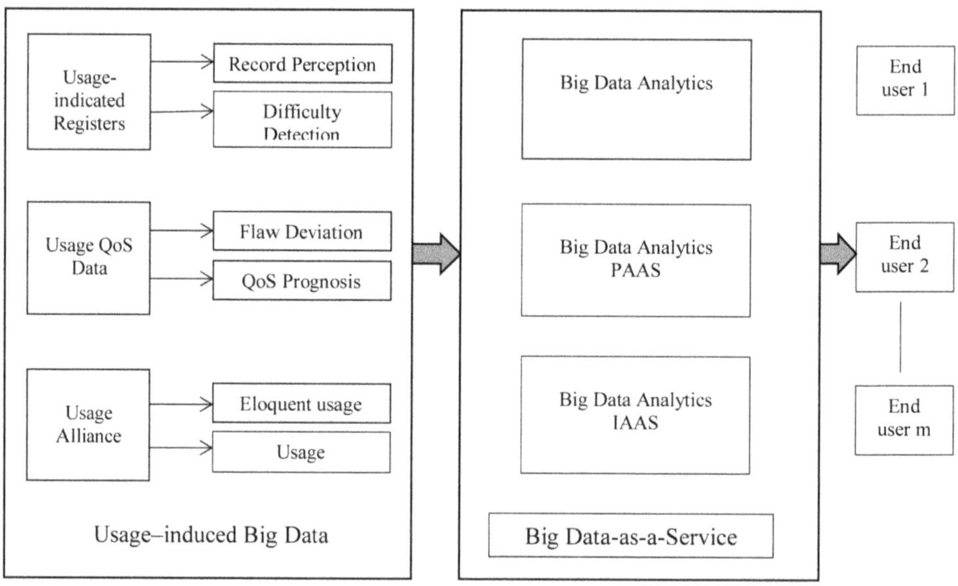

empowers dynamic joining of diverse huge information and combination of various huge information examination ways to deal with make an incentive from the administration created huge information.

USAGE-INDUCED BIG DATA

These days, there are a wide range of online administrations gave on the Internet and every day utilized by a huge number of clients. Each time when you play out an inquiry, send an Email, post a micro blog or, on the other hand shop on online business Websites, you are creating a follow of information to the administrations.

As appeared in Figure 2, as the quantity of administrations and clients scales up, the administration produced information (counting administration logs, benefit QoS data, and administration relationship) are expanding, prompting the huge information marvel. With the expanding volume of administration produced huge information, how to make values from the information turns into a critical research issue. The accompanying sub-areas will depict in detail how the benefit produced information can be prepared and broke down to upgrade framework execution.

A. Usage-Indicated Registers

With the promotion of expansive scale benefit situated frameworks, also, the number expanding of administration clients (e.g., PCs, cell phones, and so on.), a gigantic volume of follow logs are produced by the administration arranged frameworks every day. There are billions of every day logs, log documents, and organized/unstructured information from a wide assortment of administration frameworks. For instance, an

Figure 2. Usage-induced Big Data

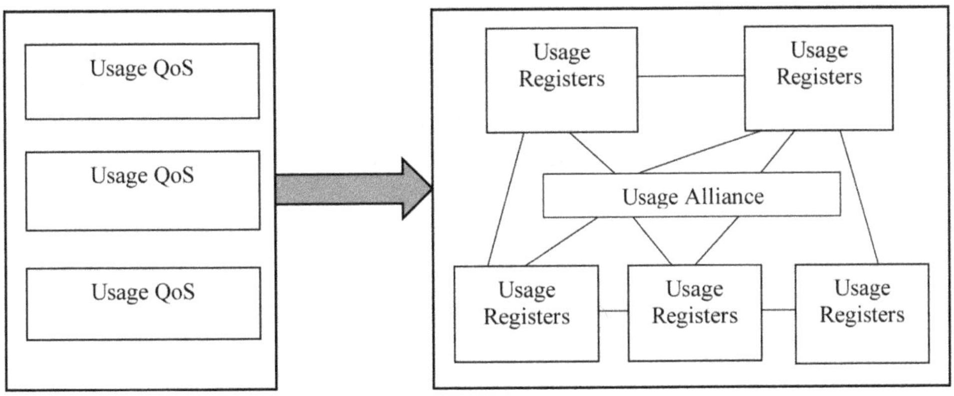

Email benefit given by Alibaba (one of the greatest web based business organization on the planet) would deliver around 30-50 gigabytes (around 120-200 million lines) of following logs every hour (H. Mi, H.Wang, Y. Zhou, M. R. Lyu, & H. Cai). These logs can be used both in the improvement stages and in ordinary operations for comprehension and investigating the conduct of the perplexing framework.

As the scale and multifaceted nature of disseminated frameworks quickly increment, extensive scale disseminated frameworks like distributed computing frameworks ordinarily include countless between benefit segments, now and again crosswise over many machines, which makes it extremely hard to physically analyze the execution issues. By the method for mining these usage-created follow logs, important data can be acquired to help benefit originators and engineers comprehend and move forward the nature of frameworks. For instance, execution bottleneck confinement can be accomplished by breaking down the follow logs. In any case, there are as yet many difficulties to be tended to. On one hand, the enormous volume of administration produced follow logs make execution finding work concentrated; then again, the interest for continuous framework analysis is continually expanding.

This area talks about how to research the follow logs to discover the esteem covered up in it, including follow log perception furthermore, execution issue conclusion.

1. **Induced Register Envision:** Disseminated frameworks are constantly developing in scale and administration segment collaborations. It is getting to be noticeably troublesome for the framework planners and chairmen to comprehend the attributes of framework execution. Ask for following methodologies, which are broadly embraced by organizations, for example, Google (B. H. Sigelman, L. A. Barroso, M. Burrows, P. Stephenson, M. Plakal, D. Beaver, S. Jaspan, & C. Shanbhag), Microsoft (S. Han, Y. Dang, S. Ge, D. Zhang, & T. Xie) and eBay (M. Y. Chen, A. Accardi, E. Kiciman, J. Lloyd, D. Patterson, A. Fox, & E. Brewer), record the execution data of solicitations (e.g., entering also, leaving time when individual solicitations experience benefit parts). These follow logs contain a great deal of shrouded fortunes for framework architects and overseers. Log representation gives apparatuses to digest representation of log records (or results of log inquiries) in a way that can be best comprehended by clients (C. Lim, N. Singh, & S. Yajnik). With the expanding volume of follow logs, effective log representation of turn into a test examine issue.

To address this issue, various methodologies have been proposed. Stardust (E. Thereska, B. Salmon, J. Strunk, M. Wachs, M. Abd-El-Malek, J. Lopez, & G. R. Ganger) utilizes social databases as store which endures poor inquiry effectiveness in the earth of huge information volumes. P-tracer (H. Mi, H. Wang, H. Cai, Y.

Zhou, M. R. Lyu, & Z. Chen) is an online execution profiling device to pictures multi-dimensional measurable data to help chairmen comprehend the framework execution practices top to bottom. DTrace (B. M. Cantrill, M. W. Shapiro, & A. H. Leventhal) and gprof (S. L. Graham, P. B. Kessler, & M. K. Mckusick) picture the execution of frameworks as call charts to connote where demands invest energy.

In spite of the fact that various past research examinations have been directed at administration log representation, this exploration issue turns out to be all the more difficult in the situation of enormous information, created by the quick increment of log documents, the unstructured log information, and the prerequisite of continuous inquiry and show. More research examinations are expected to empower constant handling and representation of the enormous volume of follow logs.

2. **Detection of Difficulties in Functioning:** In today's disseminated frameworks, particularly the cloud frameworks, an administration demand will experience diverse hosts, conjuring various programming modules. At the point when the administration can't fulfill the guaranteed benefit level understanding (SLA) to clients, it is basic to distinguish which module (e.g., a summoned technique) is the underlying driver of the execution issue in an opportune way. Follow logs give important data to discover the reason for execution issues. Step by step instructions to misuse the gigantic follow logs successfully what's more, effectively to help creator comprehend framework execution what's more, find execution practices turns into a pressing and testing research issue.

In late writing, an expansive assortment of research work has been researched to address this issue. For instance, Magpie (P. Barham, A. Donnelly, R. Isaacs, & R. Mortier), Pip (P. Reynolds, C. Killian, J. L. Wiener, J. C. Mogul, M. A. Shah, & A. Vahdat) and Iron model (E. Thereska & G. R. Ganger) are application-particular approaches, which require space learning to build the execution conclusion models. Pinpoint (M. Chen, E. Kiciman, E. Fratkin, A. Fox, & E. Brewer) utilizes bunching calculation to gathering disappointment and achievement logs. Spruce (B. H. Sigelman, L. A. Barroso, M. Burrows, P. Stephenson, M. Plakal, D. Beaver, S. Jaspan, & C. Shanbhag) utilizes Bigtable to deal with the extensive volume of follow logs. Information mining innovations, for example, chief part examination (PCA) and powerful chief part investigation are additionally utilized for distinguishing execution bottlenecks through demonstrating and mining the follow logs (H. Mi, H.Wang, Y. Zhou, M. R. Lyu, & H. Cai), (L. Shao, J. Zhang, Y. Wei, J. Zhao, B. Xie, & H. Mei, 2007).

In any case, the vast volume and high speed of usage-created follow logs make it extremely hard to perform real-time analysis. The vast majority of the past arrangements experience the ill effects of low proficiency in dealing with expansive volume of information. More effective capacity, administration, and investigation approaches for usage-created follow logs are required.

B. Usage QoS Data

The present expansive scale dispersed stages (e.g., different cloud stages) give various administrations to heterogeneous also, expanded clients. Extensive volume of QoS information of these administrations are recorded, in both server-side and client side. Since distinctive clients may watch very unique QoS execution (e.g., reaction time) on a similar administration, the volume of user-side QoS information is considerably bigger than that of server-side QoS information. In addition, QoS estimations of administration segments are evolving progressively every once in a while, bringing about touchy increment of client side administration QoS data.

Profitable data can be acquired through researching these client side administration QoS data keeping in mind the end goal to upgrade framework execution, for instance, to accomplish versatile blame resilience and to make customized QoS expectation for clients.

1. **Flaw Deviation:** The administration figuring condition is exceedingly unique and heterogeneous, where unique administrations might be incapacitated, new administrations might be included, and QoS of the administrations may change every now and then. Building solid benefit situated frameworks is substantially more difficult in this exceedingly powerful condition contrasted and the conventional remain solitary programming frameworks. Programming adaptation to internal failure (M. R. Lyu) is a vital way to deal with construct dependable frameworks by means of utilizing practically comparable parts to endure deficiencies. On the Web, the practically identical Web administrations gave by various associations can be utilized to manufacture fault-tolerant benefit arranged frameworks. When planning adaptation to internal failure techniques, benefit QoS data can be considered to improve the execution.

The colossal number of administrations in the expansive scale circulated stages is observed ceaselessly at runtime. Extensive volume of QoS information (e.g., reaction time, accessibility, throughput, and so forth.), is recorded. In addition, subsequent to leading administration summon, the clients likewise record the QoS of the summoned administrations. How to proficiently and viably handle these extensive volume of benefit QoS information to configuration adaptation to internal failure systems which can adjust to the dynamic condition for ideal execution is a testing research issue.

In our past work (Z. Zheng & M. R. Lyu), a preparatory middleware has been intended for blame tolerant Web administrations. Be that as it may, this middleware did not give a customized adaptation to internal failure system for various clients. In the dynamic Internet condition, server-side adaptation to non-critical failure is insufficient since the correspondence associations can bomb effortlessly. Customized user-side adaptation to non-critical failure should be considered. In addition, to speed up the investigation and calculation of the huge volume of benefit QoS data, web based learning calculations (H. Yang, Z. Xu, I. King, & M. Lyu) will should be explored for incremental refresh of the blame resistance procedure when new QoS values end up plainly accessible.

2. **QoS Prognosis:** Web benefit QoS forecast goes for giving customized QoS esteem forecast to administration clients, by utilizing the verifiable QoS estimations of various clients. Web benefit QoS forecast more often than excludes a client benefit framework, where every passage in the network speaks to the estimation of a specific QoS property (e.g., reaction time) of a Web benefit seen by an administration client. The client benefit lattice is normally exceptionally inadequate with many missing sections, since an administration client normally just conjured few Web benefits in the past. The issue is the means by which to precisely foresee the missing QoS values in the client benefit lattice by utilizing the accessible QoS values. Subsequent to foreseeing the missing Web benefit QoS values in the client benefit framework, each administration client can have a QoS assessment on every one of the administrations, even on the unused administrations. Thus, ideal administration can be chosen for clients to accomplish great execution.

In administration processing, Web benefit QoS expectation has pulled in a considerable measure of consideration as of late. Various QoS expectation approaches have been proposed to address this paper, counting client based QoS forecast approach (L. Shao, J. Zhang, Y. Wei, J. Zhao, B. Xie, & H. Mei), mix of client based and thing based methodologies (Z. Zheng, H. Ma, M. R. Lyu, & I. King), ranking-oriented approach, (Z. Zheng, H. Ma, & M. R. Lyu,) and I. King bunching based approach (X. Chen, Z. Zheng, X. Liu, Z. Huang, & H. Sun), and so on. Some current work (M. Tang, Y. Jiang, J. Liu, & X. F. Liu), (W. Lo, J. Yin, S. Deng, Y. Li, & Z. Wu) additionally considers the areas of Web administrations and administration clients to improve the unique situation data of the administration condition and accordingly make strides the expectation quality. Nonetheless, with the quick expanding of Web administrations and clients, the measure of client administration network is getting to be plainly bigger and bigger, which makes it not productive for continuous expectation. Since the Internet condition is exceptionally dynamic, administrations may include or drop at whatever time, so improving the vigor of QoS expectation approaches (e.g., settling the frosty begin issue) is exceptionally basic. What's more, estimations of some client side QoS properties (e.g., reaction time) are changing after some time, making the lattice turn into a three-dimensional user service-time network. Along these lines, how to effectively prepare the huge volume of accessible administration QoS information and precisely anticipate the missing QoS values in the colossal client benefit time lattice turns into an exceptionally difficult research issue.

C. Usage Alliance

These days, substantial scale dispersed frameworks regularly include a substantial number of administration segments. These administration segments are commonly conveyed in disseminated PC hubs (i.e., physical machines or virtual machines) and have complex conjuring connections. For instance, to produce the dynamic Web content for a page in one of the internet business destinations in Amazon, each demand regularly requires the page rendering segments to build its reaction by sending solicitations to more than 150 administrations (G. DeCandia, D. Hastorun, M. Jampani, G. Kakulapati, A. Lakshman, A. Pilchin, S. Sivasubramanian, P. Vosshall, & W. Vogels). These segments are related between each other. The conjuring relationship among administration parts can be displayed as a weighted coordinated diagram, where a hub in the chart speaks to an administration segment also, a guided edge starting with one hub then onto the next speaks to a part summon relationship. The weight at each edge might be communicated as the cost or the recurrence of the conjuring. This administration summon chart can be refreshed progressively at runtime, brought on by reasons, for example, benefit include/drop, benefit relocation, stack adjust, and so on.

By misusing the administration summon chart, important data can be gotten to distinguish noteworthy administration parts what's more, to empower better powerful administration relocation.

1. **Recognition of Eloquent Usage Constituent:** Unwavering quality of various administration segments may force distinctive impacts on the administration situated framework. By examining the benefit summon chart, the critical administration segments (e.g., center parts or feeble segments) can be recognized. This can significantly help us see how to enhance the structure of a framework and how to enhance the unwavering quality of the framework. For instance, extra adaptation to non-critical failure techniques can be intended for these huge administration parts to accomplish higher unwavering quality. Nonetheless, because of the way of dynamic creation of administration segments, the administration conjuring chart can be ceaselessly refreshed at runtime. Also,

the pattern towards huge scale frameworks make the administration summon chart very substantial and complex. Stochastic positioning procedures can be utilized to distinguish the critical administration part in the chart for an appropriated framework.

In our preparatory examination, we considered a segment huge on the off chance that it is summoned by numerous other imperative segments as often as possible, and proposed an irregular walk based way to deal with recognize critical segments in a cloud framework. Plus, Liu et al. (A. Liu, Q. Li, L. Huang, & S. Wen) propose to misuse the administration relationship to help the notoriety calculation of administrations, which additionally enhance the strength of the framework. Be that as it may, there are still a ton of research issues to be tended to, for instance: (1) demonstrating the connection between various benefit parts (e.g., segments an and b conjure the same part c, then there is understood connection between part a and b; (2) demonstrating the effect of segment on the entire framework, which can help us enhance the heartiness of entire framework; (3) planning more proficient and successful ways to deal with fabricate and dissect the administration conjuring chart what's more, distinguish huge administration parts.

2. **Usage Expatriation:** Dispersed frameworks ordinarily incorporate various administration segments, which should be sent to circulated hubs (i.e., physical machines or virtual machines). Since the administration condition is very powerful, after the underlying sending, it is basic to enhance the benefit sending methodology among hopeful hubs intermittently to accomplish ideal general framework execution while limiting the operational cost. Accordingly, dynamic administration movement is in need by moving the administration from one physical machine to another at runtime. It is a typical practice in numerous business cloud stages.

By demonstrating and misusing the administration summon relationship what's more, past administration use encounters, a legitimate movement of the administrations can enhance the experience for existing clients. In our past work (Y. Kang, Z. Zheng, & M. Lyu), a preparatory model has been defined in light of whole number programming to make an ideal redeployment of administrations. In any case, this number programming based model experiences the scaling issue when confronting countless and competitor hubs. This model is additionally enhanced in (J. Zhu, Z. Zheng, Y. Zhou, & M. R. Lyu) by taking administration relationship into account, which proposes to utilize a hereditary calculation to settle the demonstrate productively. With the expansion of cloud organization, some other work (e.g., (Q. Zhang, Q. Zhu, M. F. Zhani, & R. Boutaba; M. Steiner, B. G. Gaglianello, V. K. Gurbani, V. Hilt,

W. D. Roome, M. Scharf, & T. Voith, M. Alicherry & T. V. Lakshman) additionally considers the dynamic administration arrangement and relocation in geologically appropriated mists to get ideal the administration execution.

To adapt to the developing size of the administration movement issue, more productive methodologies are required. For instance, (1) notwithstanding considering the conjuring relationship among administrations, areas of administration clients can be considered to additionally enhance the movement execution; (2) utilizing QoS expectation procedures, arrange latencies among administration segments and amongst clients and administrations can be anticipated to accomplish better administration movement execution; (3) to speed up the calculation of ideal administration relocation procedure, of number programming, different machine learning calculations, for example, web based learning, and k-middle enhancement show, hereditary calculation, and so on., can be examined to empower proficient benefit movement.

CONCLUSION AND FUTURE WORK

This paper gives a diagram of administration produced huge information and Big Data-as-a-Service. Three sorts of usage-created huge information are misused to upgrade nature of service-oriented frameworks. To give basic usefulness of enormous information administration and examination, Big Data-as-a-Service is explored to give APIs to clients to get to the administration created enormous information and the huge information examination comes about.

Later on, next to the administration follow logs, QoS data what's more, administration relationship, more sorts of administration created huge information will be researched. More complete investigations of different administration produced huge information investigation methodologies will be led. Point by point innovation guide will be given what's more, security issues past the extent of this paper will likewise be explored.

REFERENCES

Alicherry, M., & Lakshman, T. V. (2012). Network aware resource allocation in distributed clouds. *Proceedings - IEEE INFOCOM*, *12*, 963–971.

Austin. (2012). *eDiscovery Trends: CGOCs Information Lifecycle Governance Leader Reference Guide*. Retrieved from http://www.ediscoverydaily.com

Barham, P., Donnelly, A., Isaacs, R., & Mortier, R. (2004). Using magpie for request extraction and workload modelling. *Proceedings of the 6th conference on Symposium on Opearting Systems Design & Implementation (OSDI'04)*, 18–18.

Beyer & Laney. (2012). *The importance of 'big data': A definition.* Gartner.

Cantrill, B. M., Shapiro, M. W., & Leventhal, A. H. (2004). Dynamic instrumentation of production systems. *USENIX Annual Technical Conference*, 15–28.

Chen, M., Kiciman, E., Fratkin, E., Fox, A., & Brewer, E. (n.d.). Pinpoint: problem determination in large, dynamic internet services. *Proceedings of the International Conference on Dependable Systems and Networks (DSN'02)*, 595–604. 10.1109/DSN.2002.1029005

Chen, M. Y., Accardi, A., Kiciman, E., Lloyd, J., Patterson, D., Fox, A., & Brewer, E. (2004). Path-based faliure and evolution management. In *Proceedings of the 1st conference on Symposium on Networked Systems Design and Implementation.* USENIX Association.

Chen, X., Zheng, Z., Liu, X., Huang, Z., & Sun, H. (2011). *Personalized QoS-aware Web service recommendation and visualization. IEEE Transactions on Services Computing.*

DeCandia, Hastorun, Jampani, Kakulapati, Lakshman, Pilchin, … Vogels. (2007). Dynamo: amazon's highly available key-value store. *Proc. 21st ACM Symposium on Operating Systems Principles (SOSP'07)*, 205–220.

Graham, S. L., Kessler, P. B., & Mckusick, M. K. (1982). Gprof: A call graph execution profiler. *ACM SIGPLAN Notices, 17*(6), 120–126. doi:10.1145/872726.806987

Han, S., Dang, Y., Ge, S., Zhang, D., & Xie, T. (2012). Performance debugging in the large via mining millions of stack traces. *Proc. 34th Int'l Conf. on Software Engineering (ICSE'12)*, 145–155. 10.1109/ICSE.2012.6227198

IBM. (2013). *What is big data? ł bringing big data to the enterprise.* Retrieved from http://www-01.ibm.com/software/data/bigdata

Kang, Y., Zheng, Z., & Lyu, M. (2012). A latency-aware co-deployment mechanism for cloud-based services. *Proceedings of the IEEE 5th International Conference on Cloud Computing (CLOUD'12)*, 630–637. 10.1109/CLOUD.2012.90

Lim, F., Singh, N., & Yajnik, S. (2008). A log mining approach to failure analysis of enterprise telephony systems. *Proceedings of the IEEE International Conference on Dependable Systems and Networks (DSN'08)*, 398–403.

Liu, Li, Huang, & Wen. (2012). Shapley value based impression propagation for reputation management in web service composition. *Pro. IEEE 19th Int'l Conf' on Web Services (ICWS'12)*, 58–65.

Lo, W., Yin, J., Deng, S., Li, Y., & Wu, Z. (2012). Collaborative web service qos prediction with location-based regularization. *Pro. IEEE 19th Int'l Conf' on Web Services (ICWS'12)*, 464–471. 10.1109/ICWS.2012.49

Lyu, M. R. (1995). *Software Fault Tolerance. In Trends in Software*. Wiley.

Mi, H., Wang, H., Cai, H., Zhou, Y., Lyu, M. R., & Chen, Z. (2012). Ptracer: Path-based performance profiling in cloud computing systems. In *Proceedings of the 36th IEEE Annual Computer Software and Applications Conference (COMPSAC'12)*. IEEE.

Mi, Wang, & Zhou, Lyu, & Cai. (2013). Towards fine-grained, unsupervised, scalable performance diagnosis for production cloud computing systems. *IEEE Transactions on Parallel and Distributed Systems*.

Reynolds, P., Killian, C., Wiener, J. L., Mogul, J. C., Shah, M. A., & Vahdat, A. (2006). Pip: detecting the unexpected in distributed systems. *Proceedings of the 3rd conference on Networked Systems Design & Implementation (NSDI'06)*, 9–9.

Shao, L., Zhang, J., Wei, Y., Zhao, J., Xie, B., & Mei, H. (2007). Personalized QoS prediction for Web services via collaborative filtering. *Proc. 5th Int'l Conf. Web Services (ICWS'07)*, 439–446. 10.1109/ICWS.2007.140

Sigelman, Barroso, Burrows, Stephenson, Plakal, Beaver, … Shanbhag. (2010). *Dapper, a large-scale distributed systems tracing infrastructure*. Google, Inc.

Steiner, M., Gaglianello, B. G., Gurbani, V. K., Hilt, V., Roome, W. D., Scharf, M., & Voith, T. (2012). Network-aware service placement in a distributed cloud environment. *Proc. ACM SIGCOMM*'12, 73–74. 10.1145/2342356.2342366

Tang, M., Jiang, Y., Liu, J., & Liu, X. F. (2012). Location-aware collaborative filtering for qos-based service recommendation. *Pro. IEEE 19th Int'l Conf' on Web Services (ICWS'12)*, 202–209.

The Economist. (2010). A special report on managing information: Data, data everywhere. *The Economist*.

Thereska, G., Salmon, B., Strunk, J., Wachs, M., Abd-El-Malek, M., Lopez, J., & Ganger, G. R. (2006). Stardust: tracking activity in a distributed storage system. ACM SIGMETRICS Performance Evaluation Review, 34(1), 3–14. doi:10.1145/1140277.1140280

Thereska, H., & Ganger, G. R. (2008). Ironmodel: robust performance models in the wild. *Proceedings of the International Conference on Measurement and Modeling of Computer Systems (SIGMETRICS '08)*, 253–264. 10.1145/1375457.1375486

Xu, W., Huang, L., Fox, A., Patterson, D., & Jordan, M. I. (2009). Detecting large-scale system problems by mining console logs. *Proceedings of the ACM 22nd Simposium on Operating Systems Principles (SOSP'09)*, 117–132. 10.1145/1629575.1629587

Yang, H., Xu, Z., King, I., & Lyu, M. (2010). Online learning for group lasso. *International Conference on Machine Learning (ICML'10).*

Zhang, Q., Zhu, Q., Zhani, M. F., & Boutaba, R. (2012). Dynamic service placement in geographically distributed clouds. *Proc. IEEE 32nd Int'l Conf. on Distributed Computing Systems (ICDCS'12)*, 526–535. 10.1109/ICDCS.2012.74

Zheng, Z., & Lyu, M. R. (2009). A QoS-aware fault tolerant middleware for dependable service composition. *Proc. 39th Int'l Conf. Dependable Systems and Networks (DSN'09)*, 239–248. 10.1109/DSN.2009.5270332

Zheng, Z., Ma, H., Lyu, M. R., & King, I. (2011). QoS-aware Web service recommendation by collaborative filtering. *IEEE Transactions on Services Computing*, *4*(2), 140–152. doi:10.1109/TSC.2010.52

Zheng, Z., Zhang, Y., & Lyu, M. R. (2010). CloudRank: A QoS-driven component ranking framework for cloud computing. *Proc. Int'l Symp. Reliable Distributed Systems (SRDS'10)*, 184–193. 10.1109/SRDS.2010.29

Zheng, Z., Zhu, J., & Lyu, M. R. (2013). Usage-created Big Data and Big Data-as-a-Service: An Overview. *IEEE International Congress on Big Data.*

Zhu, J., Zheng, Z., Zhou, Y., & Lyu, M. R. (2013). Scaling service-oriented applications into geo-distributed clouds. *Pro. IEEE Int'l Workshop on Internet-based Virtual Computing Environment (iVCE'13).*

Chapter 5
Glorified Secure Search Schema Over Encrypted Secure Cloud Storage With a Hierarchical Clustering Computation

Shweta Annasaheb Shinde
VIT University, India

Prabu Sevugan
VIT University, India

ABSTRACT

This chapter improves the SE scheme to grasp these contest difficulties. In the development, prototypical, hierarchical clustering technique is intended to lead additional search semantics with a supplementary feature of making the scheme to deal with the claim for reckless cipher text search in big-scale surroundings, such situations where there is a huge amount of data. Least relevance of threshold is considered for clustering the cloud document with hierarchical approach, and it divides the clusters into sub-clusters until the last cluster is reached. This method may affect the linear computational complexity versus the exponential growth of group of documents. To authenticate the validity for search, minimum hash sub tree is also implemented. This chapter focuses on fetching of cloud data of a subcontracted encrypted information deprived of loss of idea and of security and privacy by transmission attribute key to the information. In the next level, the typical is improved with a multilevel conviction privacy preserving scheme.

DOI: 10.4018/978-1-5225-3643-7.ch005

INTRODUCTION

Individuals are profited with cloud computing as cloud computing reduces it work and make computing and storage simplified. (Liang, Cai, Huang, Shen & Peng, 2012), (Mahmoud & Shen, 2012), (Shen, Liang, Shen, Lin & Lou, 2012). Data can be stored remotely in the cloud server as data outsourcing and accessed publicly. This embodies a mountable, constant and low-cost method for public access of data as per the high productivity and mount ability of cloud servers, and so it is favored.

Sensitive privacy information is of concern. Data should be encrypted before sending to the cloud servers (Jung, Mao, Li, Tang, Gong & Zhang, 2013), (Yang, Li, Liu & M, 2014). The data encryption comes with it the difficulty of searching the data on the cloud servers. (Cao, Wang, Li,Ren & Lou, 2014) Encryption comes with it many of other security apprehensions. Secure Sockets Layer is used by Google search to encrypt the connection be the authors the google server and search user.

Nevertheless, if the user clicks from the authors site of search result, to another the authors site will identify the search terms the user has used.

On dealing with the above matters, the searchable form of encryption (e.g., (Song,Wagner & Perrig, 2000), (Li,Xu,Kang,Yow & Xu, 2014), (Li,Lui,Dai,Luan & Shen, 2014)) has been established as a basic method to allow searching over encrypted data of cloud, which profits the procedures. At first the owner of data will produce quite a few keywords rendering to the outsourced data. Cloud server will be used to store this encrypted keywords. When the outsourced data needs to be accessed, it can choice approximately appropriate keywords and direct the cipher text of the designated keywords to the cloud server. The cloud server then usages the cipher text to contest the outsourced keywords which are encrypted, and finally will yields the matching consequences to the user who search. To attain the like search effectiveness and accuracy over data which is encrypted as like plaintext search of keyword, a widespread form of research has been advanced in literature. Wang et al.(2014) recommended a ranked keyword search system which deliberates the scores of relevance's of keywords. Inappropriately, because of using order-preserving encryption (OPE)(Boldyreva,Chenette, Lee & Oneill, 2009) to attain the property of ranking, the planned arrangement cannot attain unlikability of trapdoor.

Later, Sun et al.(Sun,Wang,Cao,Li,Lou,Hou & Li, 2013)suggested a multi-keyword text search arrangement which deliberates the scores of relevance's of the keywords and exploits a multidimensional tree method to realize the authors organized query of search. (J. Yu, P. Lu, Y. Zhu, G. Xue, & M. Li, 2013)suggested a multi-keyword top-k retrieval organization which practices fully homomorphic encryption to encrypt the index/trapdoor and assurances high security. Cao et al. (2014) suggested a multi-keyword ranked search (MRSE), which put on machine of coordinate as the matching of keyword rule, i.e., it will return the data with the maximum matching of keywords. Even though many of the functionalities of search have been advanced in former literature on the way to exact and the authors organized searchable encryption, it is still problematic for searchable encryption to attain the similar user involvement as that of the plaintext search, like Google search. This mostly attributes to subsequent two issues. At first, query with user favorites is popular in the search of plaintext (Liang, Cai,Huang,Shen, & Peng, 2012), (Mahmoud & Shen, 2012). It allows tailored search and can more precisely represent requirements of users, but has not been methodically studied and maintained in the encrypted domain of data. At second, to further improve the user's experience on searching, a significant and vital function is to allow the multi-keyword search with the comprehensive logic operations, i.e., the "AND", "OR" and "NO" operations of keywords. This is vital for search users to trim the space of searching and rapidly classify the anticipated data.

Cao et al. advise the coordinate matching search scheme (MRSE) which can be the authors as a searchable encryption system with "OR" operation (Shen,Liang, Shen,Lin, & Luo, 2014) recommended a conjunctive keyword search scheme which can be observed as a searchable encryption scheme with "AND" operation with the refunded documents matching all keywords. Though, most current suggestions can only allow search with single logic operation, somewhat than the mixture of numerous logic operations on keywords, which encourages the work.

Here, the authors discourse above two issues by emerging two Fine-grained Multi-Keyword Search (FMS) arrangements over encrypted data of cloud. Our unique donations can be abridged in three characteristics as tracks:

- The authors familiarize the relevance scores of relevance's and the preference factors of keywords for searchable encryption. The scores of relevance's of keywords can allow more detailed refunded consequences, and the factors of preferences of keywords signify the standing of keywords in the search for keyword set quantified by search of users and consistently the authors search at personalized level to cater to precise preferences of users. Thus, additional advances the search of functionalities and experience of user.
- The authors understand the "AND", "OR" and "NO" processes in the multi-keyword search for searchable encryption. Associated with arrangements in the proposed arrangement can accomplish more all-inclusive functionality and the authors query complexity of query.
- The authors employment the classified sub-dictionaries technique to improve the effectiveness of the above two arrangements. Extensive experimentations establish that the enhanced arrangements can attain better effectiveness in terms of building of index, trapdoor generating and query in the judgement with arrangements in.

Hardware restriction of mobile devices is overcome by Mobile cloud computing (Dinh, Lee, Niyato, & Wang, 2013; Li, Dai, Tian, & Yang, 2009) by discovering the accessible and virtualized storage of cloud and computing resources, and consequently can deliver much more significant and mountable services of mobile to user. In the technology of mobile cloud computing, mobile operators characteristically are outsourcing their information to cloud servers which are external, e.g., iCloud, to adore a steady, low-cost and climbable way for storage of data and access. Though, as outsourced data has sensitive private information, such as personal photos, emails., which would lead to severe confidentiality and privacy violations (W. Sun, 2013), if without efficient protections. It is therefore essential to encrypt the sensitive data before outsourcing them to the cloud. The data encryption, the authors would result in salient problems when other users need to access interested data with search, due to the problems of search over encrypted data. This fundamental issue in mobile cloud computing consequently inspires an extensive body of investigation in the recent years on the examination of searchable encryption performance to attain the authors ll-organized thorough over outsourced encrypted data (Wang, Lou, & Hou, 2014; Yang, Liu, & Yang, 2014)

An assortment of research the whole thing have freshly been established about multi-keyword search over the data which is encrypted. Cash et al. (Jarecki,Jutla,Krawczyk,Ro^{3}u, & Steiner, 2013) offer a symmetric searchable encryption organization which attains high effectiveness for big databases with uncertain on security guarantees. Cao et al. (Cao, Wang, Li, Ren, & Lou, 2014)suggest a multi-keyword search structure supportive consequence ranking by approving *k*-nearest neighbors (kNN)technique (Wong, Cheung,Kao, & Mamoulis,2009). Naveed et al. (2014) proposition an active searchable encryption system complete blind storage to obscure admittance pattern of the user for search. In demand to encounter the practical search necessities, search concluded data which is encrypted should provision the subsequent three functions. First, the encryption systems which are searchable should provision multi-keyword search, and deliver the same experience for user as thorough in search for Google with different keywords; search for single-keyword is far from acceptable by only recurring very incomplete and imprecise results for results. Second, to rapidly classify most applicable results, the user for search would characteristically favor cloud servers to category the refunded search consequences in a relevance-based command (Pang, Shen, & Krishnan, 2010) ranked by the order of relevance of the request for search documents. In accumulation, display the search based on rank to users can also eradicate the needless traffic of network by only distributing back the utmost results which are relevant from to search users from cloud. Third, as for the effectiveness of search, then the quantity of the documents which are imperfect in a database could be tremendously large, encryption which is searchable constructions should be organized in the authors-

mannered to quickly response to the requirements for search with interruptions and they are smallest. In modification to the proposed prosperities, frequently of the usual proposals, the authors, nose dive to proposal satisfactory intuitions near the construction of full performed encryption which is searchable. As an application near the subject, the authors proposition and the authors-organized multi-keyword ranked search (EMRS) preparation over encrypted cloud data for mobile through blind storage. Our important charities can be abridged as surveys:

- The authors explain a relevance for score in encryption which is searchable to accomplish multi-keyword ranked search finished the cloud data for mobile which is encrypted. In gathering to that, the authors proposition and the authors organized index to advance the efficiency for search.
- By adapting the blind storage scheme in the Meurthe authors resolve the trapdoor unlikability problematic and obscure admittance pattern of the user for search from the cloud server.
- The authors give systematic analysis of security to prove that the EMRS can spread a high security level counting documents for confidentiality and index, privacy with trapdoor, trapdoor unlikability, and covering admission pattern of the search user. Moreover, the authors implement extensive experiments, which show that the EMRS can achieve enhanced efficiency in the terms of functionality and search effectiveness compared with prevailing proposals.

Cloud computing is known as a substitute to outdated information technology and has been progressively familiar as the greatest significant revolving point in the expansion of information technology owing to its inherent sharing of resource and low maintenance characters. Cloud computing is a computing model, where the information which is shared, resources and software are supplied to devices and computers based on requirement. This allows the end user to admittance the resources for cloud computing anytime from required platform such as mobiles, or desktops which is the mobile computing platform.

Clouds are huge pools of easily practical and available resources which are virtualized. The data and the applications which are software essential by the workers are not stored on their self-computers; in its place, they are the authors on servers which are remote which are the control of users. It is a model which is pay-per-use in that the structure benefactor by resources of service level agreements(SLAs) which (Vaquero, Rodero-Merino, Caceres, & Lindner, 2009). As cloud computing becomes prevalent, more and more sensitive information's are being centralized into the cloud. Such as emails, photo albums, personal health records, financial transactions, tax documents and government documents etc.

The detail that owners of data and cloud server are no extended in the similar trusted domain may place the data at risk outsourced unencrypted. The cloud server escape information of data to illegal enables or can be hacked. To deliver privacy for data, data which is sensitive needs to be encrypted first of outsourcing to the profitable public cloud (Kamara & Lauter, 2010). The unimportant explanation of transferring all the information and decrypting in the vicinity is clearly unreasonable, due to the gigantic amount of band width rate in level of cloud scale systems.

Discovering preserving privacy and real search over encrypted date of cloud is of supreme position seeing the possibly great amount of on claim users of data & enormous quantity of outsourced document of data in the cloud, this problem is predominantly challenging as it is tremendously difficult to meet also the necessities of presentation, system usability and scalability encryption makes operative utilization of data a very stimulating task given that there could be a big quantity of outsourced information files. Also in the cloud computing owners of data may portion their outsourced information with numerous users who might want to only recover certain exact files of data. They are engrossed in during a conference. One of the utmost prevalent imposts to do so is whole keyword search method licenses users to intelligently recuperate files of interest. Need for information retrieval is the most commonly occurring commission in cloud to the user to from server. Usually, cloud servers complete relevance result ranking in instruction to make the exploration as earlier. Such ranked search scheme allows users of data to find the most applicable info quickly, instead of returning undistinguishable results. Ranked search can stylishly remove unnecessary traffic for the network by distribution back only the greatest data for relevance which is highly wanted in the "Pay-As-You-Use" paradigm for cloud.

For confidentiality shield, such process of ranking, the authors, must not escape any keyword connected data. On the other side, to recuperate the result for accuracy of search as the authors as to recover the searching for user experience, it is also necessary for such ranking system to support multiple keywords search, as single keyword search often yields far too coarse results. As a common practice indicated by today's the authors search engines (e.g., Google search), data users may tend to provide a set of keywords instead of only one as the indicator of their search interest to retrieve the most relevant data. And each keyword in the search request can help narrow down the search result further. "Coordinate matching" (Witten,Moffat & Bell,1999), i.e., as numerous matches as conceivable, is a the authors-organized similarity amount among such multi-keyword meaning to improve the relevance of result, and has been extensively used in the plaintext data retrieval (IR) community. Though, how to put on it in the encrypted cloud data search scheme remains a very stimulating assignment because of distinguishing privacy and security problems, counting numerous strict supplies like the privacy of data, the privacy of index, privacy of keyword, and many more.

Cloud computing is known as a substitute to old-style data technology and has been progressively familiar as the greatest important rotating point in the expansion of information technology due to its inherent sharing of resource and maintenance of low characters. Cloud computing is also an internet based computing model, where the information is shared, resources and software are on condition that on other devices and computers on demand upon demand. This allows the end user to admission the cloud computing possessions anytime any platform such as a mobile computing platform, cell phone Clouds are big puddles of effortlessly serviceable and available resources which are virtualized. The information and the applications of software compulsory by the workforces are not deposited on their own processers; instead they are deposited on remote servers which will be under the control of other users. it is a pay-per- use model in which the organization earner by resources of service level agreements customized (SLAs) (Vaquero, Rodero-Merino, Caceres, & Lindner, 2009)

Security in Cloud

There are a portion of interests to tolerant Cloud Computing, there are also some significant walls to receipt (Seung Hwan, Gelogo & Park, 2012). One of the greatest major fences to acceptance is the security, surveyed by matters concerning acquiescence, privacy and matters which are authorized. Since Cloud Computing characterizes a comparatively new figuring model, there is an enormous deal of ambiguity about how security at every level (network, host, application, data levels, etc.) can be attained and in what way security of application is stimulated to Cloud Computing. That indecision has dependably controlled information managers to state that security is their number one apprehension with Cloud Computing. Security anxieties recount to hazard areas such as outside data storage, dependence on the internet which is public, absence of control, integration and multitenancy with security is internal. Associated to conservative technologies, cloud has many exact topographies, such as its countless gauge and the detail that resources going to cloud providers are completely disseminated, heterogeneous and totally virtualized. Conservative security machineries such as Identity authentication, and authorization are no longer enough for clouds in their current form. For of the cloud facilities replicas working, the working replicas, and practices cast-off to allow services for cloud. Cloud computing may present-day dissimilar dangers to suggestion than old-style IT resolutions. Unfortunately, participating safety into these explanations is frequently supposed as making them more inflexible.

Search in Encrypted Cloud Data

As Cloud Computing turn into extensive, more delicate data are being transported into cloud, such as individual health records, emails, confidential videos and images, data for business finance, documents for government, etc. As per this i.e., storage their information into the cloud, the data owners can be reassured from the problem of information storing space and preservation so to like the on- demand high brilliance storage for data service (Reddy, 2013). Though, the reality that information suppliers and cloud servers are not in the alike reliable area may put the subcontracted information at danger. By way of the cloud server can no extended be completely reliable in such an environment for cloud since of a variety of reasons, they are: the cloud server may leakage data to illegal things or it may be slashed. It tracks that delicate information characteristically can be encrypted before outsourcing for data confidentiality and fighting undesirable admissions. Though, encryption for data makes data utilization effectiveness and efficiency a very challenging task given that there could be a large amount of outsourced data files. Furthermore, in Cloud Computing, data owners/provider may share their outsourced data with many users. The individual users shall wish to only recover certain exact files of data they are absorbed in through a given conference. One among the most recognized customs is to specifically recover files complete keyword-based search as a substitute of regaining all the files which are encrypted like before which is totally unreasoning in cloud computing circumstances (Khan,Wang,Kulsoom & Ullah, 2013) Like this keyword-based search technique allows users to meaningfully regain files of awareness and has approximately valuable in search of plaintext situations, such as Google search. Miserably, encryption for information limits user's capability to perform search for keyword and subsequently makes the out-of-date plain text search methods not appropriate for Cloud Computing.

Lately, the cloud computing pattern (Mell & T. Grance, 2011) is transforming the establishments in method of effective their information mainly in the method they accumulation, admittance and process information (Paillier,1999). As a developing calculating pattern, for cloud computing, it interests many establishments to correlated potential for cloud in relations of flexibility, cost-efficiency and rid of managerial overhead. Cloud will originate valuable and delicate data about the real data matters by detecting the variable data admission designs even if is an information is encrypted (Capitani, Vimercati, Foresti, & Samarati, 2012; Williams, Sion, & Carbunar, 2008). Most often, governments characteristic their computational procedures in accrual to their data to the cloud. The advantage of cloud is that the privacy and security matters in the cloud which circumvents the productions to use those plunders. The information can be encrypted earlier subcontracting to cloud when information is highly delicate. When information is encrypted, regardless of the fundamental encryption system, it is very interesting to accomplishment any information mining responsibilities ever decrypting the information (Samanthula, Elmehdwi, & Jiang, 2014).

Cloud computing is extended fantasized dream of computing as usefulness, from cloud customs can at all store their information into cloud so to like on-demand in height excellence claims and facilities from a public lake of configuring computing possessions (Vaquero, Lodero-Merino, Caceres, & Lindner, 2009). Its countless elasticity and financial investments are inspiring both persons and initiatives to outsource their local composite information organization scheme into cloud. To shield information confidentiality and battle the authors come y*admissions in the cloud and outside, delicate information, e.g., electronic mail, private healthiness histories, snap scrapbooks, tax papers, economic communications supposed to be encrypted by information proprietors before subcontracting to the profitable public cloud (Kamara & Lauter, 2010) This takes away the old-style information use service basis on plaintext keyword searches. The unimportance of transferring all the information and decrypting nearby is obviously unreasonable, due to the enormous quantity of bandwidth price in cloud gage schemes. Furthermore, aside from removing the local storing organization, storage information into serves for cloud no drive except they can effortlessly have examined and used. Therefore, discovering preserving privacy and real service for search done encrypted data for cloud is of supreme position. Seeing the possibly big amount of on-demand information operators and enormous quantity of subcontracted information forms in cloud, this problem is predominantly stimulating as it is tremendously problematic to encounter also the necessities of presentation, scheme scalability and usability. To encounter the real information recovery essential, the big amount of forms request the server for cloud to achieve consequence relevance ranking, in its place of recurring undistinguishable consequences. Ranked search scheme allows information operators to discovery the greatest applicable data rapidly, somewhat than categorization finished every competition in the gratified group (Singhal, 2001). Search based on rank can gracefully eradicate pointless traffic for network by distribution posterior only the most applicable information, which is extremely necessary in "pay-as-you use" cloud model. For protection, of privacy such ranking process, though, should not escape any keyword associated data. Respective, to recover the effect for accurateness of search to recover the experience for user for going through, it is important for system for ranking to provision many searches based on keywords, as sole search for keyword frequently produces far too uneven consequences. As a mutual repetition designated by todays search for the trains (e.g. Search for Google), data administrators may grade to convey a standard of watchwords in its place of just exceptional as the pointer of their enthusiasm of hunt to recoup the most relevant data. What's more, exclusively watchword seek request is brilliant to help thin discouraged the outcome for pursuit extra. "Arrange coordinating" (Witten, Moffat, & Bell,1999), i.e., as various rivalries as likely, is an efficient correlation sum among such multi-watchword intending to enhance the outcome pertinence, and broadly utilized as a part of the plaintext data recovery (IR) people group. Be that as it may, how to apply it in the scrambled cloud information seek framework remains an extremely difficult assignment in view of characteristic security and protection snags, including different strict prerequisites like the information security, the list protection, the catchphrase security, and numerous others. In the writing, searchable encryption (Song, Wagner, & Perrig, 2000; Golle, Staddon, & Waters, 2004) is a useful strategy that regards scrambled information as reports and permits a client to safely seek through a solitary catchphrase and recover archives of intrigue. In any case, coordinate use of these ways to deal with the protected extensive scale cloud information usage framework would not be fundamentally reasonable, as they are created as crypto primitives and can't suit such high administration level prerequisites like framework ease of use, client look in understanding, and simple data revelation. Some current plans have been proposed to bolster Boolean watchword look (Golle, Staddon, 2004; Shen, Shi, & Waters, 2009) as an endeavor to enhance the hunt adaptability, they are as yet not sufficient to give clients adequate outcome positioning usefulness. Our initial work (Wang, Cao, Li, Ren, & Lou, 2010) has known about this issue, and given the authors for the safe positioned seek over scrambled information issue yet just for inquiries comprising of a solitary watchword. The most effective method to outline a proficient scrambled information seek system that backings multi-catchphrase semantics without security ruptures remains a testing open issue. In this chapter, surprisingly, the authors characterize and take care of the issue of multi-watchword positioned look over encoded cloud information (MRSE) while safeguarding strict framework savvy security in the distributed computing worldview. Among different multikey word semantics, the authors pick the proficient comparability measure of "facilitate coordinating", i.e., whatever number matches as would be prudent, to catch the pertinence of information reports to the pursuit question. The authors utilize "internal item comparability" i.e., the quantity of question watchwords showing up in a record, to quantitatively assess such similitude measure of that archive to the pursuit inquiry. Amid the file development, each record is related with a double vector as a subindex where each piece speaks to whether comparing catchphrase is contained in the report. The pursuit is likewise depicted as a parallel vector where each piece implies whether comparing watchword shows up in this hunt ask for, so the comparability could be precisely measured by the inward result of the question vector with the information vector. Be that as it may, straightforwardly outsourcing the information vector or the question vector will damage the file protection or the pursuit security. To meet the test of supporting such multi-catchphrase semantic without protection breaks, the authors propose an essential thought for the MRSE utilizing secure internal item calculation, which is adjusted from a safe k-closest neighbor (kNN) procedure and afterward give two fundamentally enhanced MRSE plots in a he authors ordered way to accomplish different stringent protection necessities in two danger models with expanded assault abilities. Our commitments are abridged as takes after:

1. For the first occasion when, the authors investigate the issue of multi keyword positioned seek over encoded cloud information, and build up an arrangement of strict protection necessities for such a safe cloud information usage framework.
2. The authors propose two MRSE plans considering the similitude measure of "organize coordinating" while at the same time meeting distinctive protection prerequisites in two diverse danger models.
3. Thorough examination exploring protection and effectiveness assurances of the proposed plans is given, and investigations on this present reality dataset additionally demonstrate the proposed conspires undoubtedly present low overhead on calculation and correspondence.

Distributed computing has been considered as another model of big business IT foundation, which can compose enormous asset of registering, stockpiling and applications, and the authors clients to appreciate omnipresent, helpful and on-request arrange access to a mutual pool of configurable processing assets with extraordinary effectiveness and negligible monetary overhead. Pulled in by these engaging components, both people and undertakings are roused to outsource their information to the cloud, rather than buying programming and equipment to deal with the information themselves. Regardless of the different favourable circumstances of cloud administrations, outsourcing touchy data, (for example, messages, individual the authors being records, organization back information, government reports, and so forth) to remote servers brings protection concerns. The cloud specialist organizations (CSPs) that keep the information for clients may get to clients' delicate data without approval. A general way to deal with ensuring the information secrecy is to encode the information before outsourcing. Notwithstanding, this will bring about an immense cost as far as information ease of use. For instance, the current systems on watchword based data recovery, which are generally utilized on the plaintext information, can't be specifically connected on the scrambled information. Downloading every one of the information from the cloud and unscramble locally is clearly unrealistic. Keeping in mind the end goal to address the above issue, scientists have planned some universally useful arrangements with completely homomorphic encryption (Gentry, 2009) or unmindful RAMs (Goldreich & Ostrovsky,1996). Nonetheless, these techniques are not common sense because of their high computational overhead for both the cloud separate and client. More down to earth extraordinary reason arrangements, for example, searchable encryption (SE) plans have made commitments as far as effectiveness, usefulness and security. Searchable encryption plans for the customer to store the encoded information to the cloud and execute catchphrase seek over cipher text area. Up until this point, plentiful works have been proposed under various risk models to accomplish different hunt usefulness. For example, single catchphrase inquiry, closeness look, multi-watchword Boolean pursuit, positioned seek, multi-catchphrase positioned look, and so forth. Among them, multikey word positioned look accomplishes increasingly consideration for its pragmatic relevance. As of late, some dynamic plans have been proposed to bolster embedding and erasing operations on record accumulation. These are critical fills in as it is profoundly conceivable that the information proprietors need to refresh their information on the cloud server. Be that as it may, few of the dynamic plans bolster productive multikey word positioned look.

Distributed computing is one method for processing. Here the figuring assets are shared by numerous clients. The advantages of cloud can be stretched out from individual clients to associations. The information stockpiling in cloud is one among them. The virtualization of equipment and programming assets in cloud invalidates the money related venture for owning the information stockroom and its upkeep. Many cloud stages like Google Drive, iCloud, SkyDrive, Amazon S3, Dropbox and Microsoft Azure give stockpiling administrations.

Security and protection concerns have been the major challenges in distributed computing. The equipment and programming security instruments like firewalls and so forth have been utilized by cloud supplier. These arrangements are not adequate to secure information in cloud from unapproved clients because of low level of straightforwardness (Cloud Security Alliance, 2009). Since the cloud client and the cloud supplier are in the diverse put stock in area, the outsourced information might be presented to the vulnerabilities (Cloud Security Alliance, 2009; Ren, Wang, & Wang, 2012; Brinkman, 2007). In this manner, before putting away the important information in cloud, the information should be encoded (Kamara & Lauter, 2010). Information encryption guarantees the information classification and trustworthiness. To save the information protection the authors must outline a searchable calculation that takes a shot at scrambled information (Wong,Cheung, Kao, & Mamoulis, 2009). Numerous specialists have been adding to seeking on scrambled information. The hunt systems might be single catchphrase look or multi watchword seek (C. Wang, N. Cao, J. Li, K. Ren, & W. Lou, 2010). In gigantic database, the inquiry may bring about many reports to be coordinated with watchwords. This causes trouble for a cloud client to experience all archives and have generally applicable reports. Look in view of positioning is another arrangement, wherein the reports are positioned in view of their pertinence to the watchwords (Singhal, 2001). Practical searchable encryption systems help the cloud clients particularly in pay-as-you utilize show.

The analysts consolidated the rank of archives with numerous watchword pursuit to think of proficient financially suitable searchable encryption strategies. In searchable encryption, related writing, calculation time what's more, calculation overhead is the two most as often as possible utilized parameters by the specialists in the space for dissecting the execution of their plans. Calculation time (moreover called "running time") is the period required to play out a computational procedure for instance looking a watchword, creating trapdoor and so forth. Calculation overhead is identified with CPU usage regarding asset distribution measured in time. In this examination work, the authors dissect the security issues in distributed storage and propose the authors for the same. Our commitment can be compressed as takes after:

1. Interestingly, the authors characterize the issue of secure positioned watchword seek over scrambled cloud information, and give such as the authors convention, which satisfies the protected positioned look usefulness with no pertinence score data spillage against watchword protection.
2. Exhaustive security examination demonstrated that our deviated based positioned searchable encryption plot utilizing CRSA and B-tree to be sure appreciates "as-solid as possible". security ensure contrasted with past searchable symmetric encryption (SSE) plans.
3. Broad exploratory outcomes exhibit the adequacy and productivity of the proposed arrangement.

Distributed computing has changed the way businesses approach IT, the authors them to end up noticeably more nimble, present new plans of action, offer more administrations, and trim down IT costs. Distributed computing innovations can be executed in a wide assortment of designs, under different administration and arrangement models, and can exist together with numerous advancements and programming outline strategies. The distributed computing foundation keeps on acknowledging unstable development. The authors, for security experts, the cloud displays a tremendous quandary: How would you grasp the advantages of the cloud while keeping up security controls over your associations' advantages? It turns into an issue of adjust to decide if the expanded dangers are genuinely justified regardless of the nimbleness and monetary advantages. Keeping up control over the information is foremost to cloud achievement. 10 years prior, undertaking information regularly lived in the association's physical framework, all alone servers in the association's server farm, where one could isolate touchy data in individual physical servers. Today, by virtualization and the cloud, information might be under the association's intelligent control, yet physically put away in foundation possessed and oversaw by an alternate element. This move in charge is the main reason new methodologies and systems are required to guarantee associations can keep up information security.

At the point when an outside party claims, controls, and oversees foundation and computational assets, how might you be guaranteed that business or administrative information stays private and secure, and that your association is shielded from harming information breaks? This makes cloud information security fundamental. Distributed computing the significance of Cloud Computing is expanding also, it is accepting a developing thought in the logical also, mechanical groups. The NIST (National Institute of Standards and Technology) proposed the accompanying meaning of distributed computing:

Distributed computing is a demonstrate for the authors ring advantageous, on-request arrange get to a common pool of configurable processing assets (e.g., systems, servers, stockpiling, applications, and administrations) that can be quickly provisioned and discharged with negligible administration exertion or specialist co-op communication. This cloud show advances accessibility (Mell & Grance, 2012).

The cloud enhances joint effort, nimbleness, versatility, accessibility, capacity to adjust to varieties as indicated by request, speed up advancement work, and gives potential to cost diminishment through enhanced and effective processing. Distributed computing consolidates various processing thoughts what's more, advances, for example, Service Oriented Architecture (SOA), The authors 2.0, virtualization and different advances with dependence on the Internet, supporting regular business applications online through the authors programs to fulfil the processing needs of clients, while their product and information are kept up on the servers.

Cloud Delivery Models are:

- **Private Cloud:** Cloud framework is provisioned for use by a solitary association that involves different inhabitants. Private mists might be worked on-or off-premises and are behind the organization firewall.
- **Public Cloud:** A cloud specialist co-op offers administrations to different organizations, scholastic foundations, government offices, and different associations with get to by means of the Internet.
- **Hybrid Cloud:** Hybrid mists join two cloud conveyance models that stay novel as elements, yet they are bound together by innovation that the authors information and application transportability. Cloud blasting is a case of one way undertaking that utilize cross breed mists to adjust loads amid pinnacle request periods.
- **Community Cloud:** Cloud foundation is provisioned for the selective utilization of a group of client associations with shared figuring prerequisites for example, security, strategy, and consistence.

The Service layers for these conveyance models are:

- **Infrastructure as an Administration (IaaS):** Cloud framework is the accumulation of equipment and programming that the authors the fundamental qualities of the cloud. IaaS permits clients to self-arrangement these assets to run stages and applications.
- **Platform as an Administration (PaaS):** PaaS the author clients to adjust legacy applications to a cloud situation or create cloud-mindful applications utilizing programming dialects, administrations, libraries, and other designer devices. Programming as an administration (SaaS) – Users can run applications by means of numerous gadgets on cloud foundation.

SECURITY IN CLOUD

Although there is a considerable measure of advantages to receiving Distributed computing, there are additionally some extensive hindrances to acknowledgment (Seung Hwan, Gelogo & Park. 2012). A standout amongst the most significant obstructions to selection is the security, trailed by issues in regards to consistence, protection and approved matters. Since Cloud Processing speaks to a moderately new registering model, there is an enormous arrangement of vulnerability about how security by any stretch of the imagination levels (arrange, have, application, information levels, and so forth.) can been accomplished and how application security is moved to Cloud Figuring. That vulnerability has reliably driven data administrators to express that security is their number one worry with Cloud Computing. Security concerns identify with hazard territories, for example, outside information stockpiling, reliance on general society the authors, absence of control, multitenancy what's more, incorporation with inside security. Contrasted with traditional advances, cloud has numerous elements, for example, its extraordinary scale and the reality that assets having a place with cloud suppliers are totally disseminated, heterogeneous and totally virtualized. Regular security components, for example, character, validation, and approval are no sufficiently longer for mists in their present shape. Considering the cloud benefit models utilized, the operational models, and the philosophies used to the authors cloud administrations, cloud registering may exhibit distinctive dangers to an affiliation than customary IT arrangements. Unfortunately, incorporating security into these arrangements is regularly seen as making them more inflexible.

The distributed computing research group, especially the Cloud Security Alliance, has perceived security issues in cloud. In its Top Threats to Cloud Processing Report (Ver.1.0) (Top Threats to Cloud Computing Report (Ver.1.0),2010), it recorded seven top dangers to distributed computing:

1. Mishandle and terrible utilization of distributed computing.
2. Uncertain application programming interfaces.
3. Pernicious insiders.
4. Shared innovation vulnerabilities.
5. Information misfortune or spillages.
6. Record, administration and movement capturing.
7. Obscure hazard profile.

ENCRYPTED CLOUD DATA

By doing as such i.e., putting away their information into the cloud, the information proprietors/suppliers can be eased from the authors right of information storage room and support to appreciate the on-request high fabulousness information stockpiling administration (Reddy, 2013). Notwithstanding, reality that information suppliers and cloud server are not in the comparative trusted space may put the outsourced information at hazard, as the cloud server may no longer be completely confided in such a cloud situation considering several reasons, they are: the cloud server may spill data substance to unapproved elements or it might be hacked. It takes after that delicate information ordinarily ought to be scrambled preceding outsourcing for information security and battling undesirable gets to. Information encryption makes information usage adequacy and proficiency an extremely difficult assignment given that there could be a lot of outsourced information records. Moreover, in Cloud Computing, information proprietors/supplier may impart their outsourced information to a substantial number of clients. The individual clients may be earning to just recover certain exact information records they are occupied with all through a given session. A standout amongst the most acknowledged routes is to specifically recover documents through catchphrase based hunt as an option of recovering all the scrambled documents back which is totally outlandish in distributed computing situations (Khan & Wang, 2010).

BACKGROUND

Firstly, they use the “Dormant Semantic Analysis” to uncover relationship amongst terms and reports. The inert semantic examination exploits certain higher-arrange structure in the relationship of terms with archives (“semantic structure”) and receives a diminished measurement vector space to speak to words and reports. In this manner, the relationship be the authors terms is naturally caught. Furthermore, their plan utilizes secure “k-closest neighbour (k-NN)” to accomplish secure hunt usefulness. The proposed plan could return the correct coordinating records, as the authors as the documents including the terms inactive semantically related to the question watchword. At long last, the exploratory result exhibits that their strategy is superior to the first MRSE conspire (Song & Wagner, 2000). Their procedures have various pivotal focal points. They are provably secure: they give provable mystery to encryption, as in the untrusted server can't learn anything about the plain content when just given the cipher text; they give inquiry disengagement to quests, implying that the untrusted server can't learn much else about the plaintext than the query output; they give controlled seeking, so that the untrusted server can't hunt down a subjective word without the client's approval; they likewise bolster concealed inquiries, so that the client may approach the untrusted server to hunt down a mystery word without uncovering the word to the serve (Sun, Wang, Cao, Li, Lou, Hou, & Li, 2013). They propose a tree-based list structure and different versatile techniques for multi-dimensional (MD) calculation so that the handy hunt proficiency is greatly improved than that of direct inquiry. To further improve the inquiry protection, they propose two secure file plans to meet the stringent protection prerequisites under solid danger models, i.e., known cipher text display and known foundation demonstrate. What's more, they devise a plan upon the proposed file tree structure to the authors genuineness check over the returned indexed lists. At long last, the authors show the viability and productivity of the proposed plots through broad trial assessment. (Yu, Lu, Zhu, Xue, & Li, 2013) Distributed computing has developing as a promising example for information outsourcing and astounding information administrations. Worries of touchy data on cloud conceivably causes security issues. Information encryption ensures information security to some degree, yet at the cost of traded off proficiency. Searchable symmetric encryption (SSE) permits recovery of encoded information over cloud. In this chapter, the concentration is on tending to information security issues utilizing SSE. Interestingly, the authors define the security issue from the part of comparability significance and plan vigor. They watch that server-side positioning considering request saving encryption (OPE) spills information security. To take out the spillage, the authors propose a two-round searchable encryption (TRSE) conspire that backings beat k multikey word recovery. (Wong, D, Cheung, Kao, & Mamoulis, 2009) In This chapter they talk about the general issue of secure calculation on an encoded database and propose a SCONEDB (Secure Computation ON an Encrypted Database) show, which catches the execution and security necessities. As a contextual analysis, the authors concentrate on the issue of k-closest neighbour (kNN) calculation on an encoded database. The authors build up another lopsided scalar-item protecting encryption (ASPE) that jelly a unique kind of scalar item. They utilize APSE to develop two secure plans that bolster kNN calculation on scrambled information; each of these plans is appeared to oppose down to earth assaults of an alternate foundation learning level, at an alternate overhead cost. Broad execution studies are done to assess the overhead and the proficiency of the plans (Zhang & Zhang, 2011). Since Boneh et al. proposed the thought and development of Public Key Encryption with Keyword Search (PEKS) conspire, numerous updates and expansions have been given. Conjunctive watchword inquiry is one of these expansions. A large portion of these built plans cannot understand conjunctive with subset catchphrases look work. Subset watchwords look implies that the beneficiary could inquiry the subset catchphrases of all the catchphrases implanted in the cipher text. The authors ponder the issue of conjunctive with subset watchwords seek work, talk about the disadvantages about the existed plans, and after that give out a more effective development of Public Key Encryption with Conjunctive-Subset Keywords Search (PECSK) conspire. A correlation with different plans about effectiveness will be displayed. They additionally list the security prerequisites of their plan, then give out the security investigation (Song, Wagner, & Perrig, 2000).

MAIN FOCUS OF THE AUTHORS

Issues, Controversies, Problems

Data Owner

The information proprietor subcontracts her information to the cloud for suitable and consistent data admission to the equivalent search operators. To defend the information confidentiality, the information proprietor encrypts the unique information over symmetric encryption. To recover the exploration effectiveness, the data owner makes approximately keywords for each subcontracted document. The equivalent index is then formed giving to the keywords and a secret key. Afterward, the data owner directs the encrypted brochures and the equivalent directories to the cloud, and directs the symmetric key and secret key to exploration workers.

Cloud Server

The cloud server is an in-the authors entity which supplies the encrypted brochures and equivalent indexes that are established from the information proprietor, and delivers information admittance and exploration facilities to exploration users. When an exploration operator directs a keyword access to the cloud server, it resolves re-emergence a group of equivalent brochures founded on firm processes.

Search User

A search operator enquiries the subcontracted brochures from the cloud server with subsequent three steps. First, the exploration operator accepts together the symmetric key and the secret key from the information proprietor. Second, as per the exploration keywords, the exploration operator usages the secret key to make hatch and directs it to cloud server. Last, he obtains the corresponding text collection from the cloud server and is decrypting them with the symmetric key.

RSA Algorithm

RSA is the process shaped by the modern computers to decrypt and encrypt messages. It is an asymmetric cryptographic technique. Asymmetric resources that there are two different keys. This is public key cryptography, since one of them can be prearranged to everyone. The other key must be private. In conclusion, the factors of an integer are rigid (the factoring problem). A worker of RSA kinds and then issues the produce of two big prime figures, with a supplementary rate, as public key of theirs. The factors which are prime remain secret. Anybody can use the public key to encrypt a communication, but with presently published approaches, if the public key is large, only somebody with information of the prime factors can practicably decode the communication. It is used by modern computers to encrypt and decrypt communications. It is an asymmetric cryptographic procedure. Asymmetric means that there are two different keys. This is public key cryptography, since one of them can be given to all. The other key must be kept private.

Hierarchical Clustering

It produces hierarchy of clusters.

Two approaches:

1. **Agglomerative:**

Being a bottom-up approach. Individually opinion starts in its individual cluster. And couple of clusters combine as one changes up the hierarchy.

2. **Divisive:**

It is top-down approach. Explanations starts in one cluster, and splits are achieved recursively as one transfers up the hierarchy.

SOLUTIONS AND RECOMMENDATIONS

Symmetric Key Algorithm

1. Symmetric encryption practices the similar key to mutually decrypt and encrypt
2. DES is used as symmetric key algorithm
3. They are fast and their complexity is low so they can be easily implemented
4. Secret key should be configured in all hosts

Asymmetric Key Algorithm

1. Asymmetric key uses one key to decrypt and one key to encrypt
2. RSA is one of the asymmetric key algorithm
3. It is slow then symmetric key algorithm
4. No need of configuring secret keys in all hosts

Knn Algorithm

1. Partitional clustering
2. Partitions independent of each other
3. Sensitive to cluster center initialization
4. Poor convergence speed and bad overall clustering can happen due to poor initialization
5. Works only for around shapes
6. Doesn't work well for non-convex shapes

Hierarchical Clustering Algorithm

1. Hierarchical clustering
2. Visualize using a tree structure
3. Can give different partitioning
4. Doesn't need specification of number of clusters
5. It can be slow
6. Two types agglomerative and divisive

FUTURE RESEARCH DIRECTIONS

The authors propose to additionally spread the application to deliberate the extensibility of the set of file and the multi-user environments of cloud. This way, some initial consequences on the extensibility and the multiuser cloud environments. Another stimulating theme is to progress the highly ascendable searchable encryption to allow the authors organized exploration on large practical databases.

CONCLUSION

Fine-grained multikey word search (FMS) method was studied over encrypted data from cloud and delivered two FMS schemes. The FMS I constitute both the preference factors and relevance scores of keywords to improve more accurate search and improved users' knowledge, correspondingly. The FMS II attains safe and the authors organized search with functionality, i.e., "AND", "OR" and "NO" operations of keywords. Additionally, the authors have projected the improved methods supporting classified sub-dictionaries (FMSCS) to advance efficiency. The authors have used hierarchical clustering computation for the same.

REFERENCES

Abdalla, M., Bellare, M., Catalano, D., Kiltz, E., Kohno, T., Lange, T., ... Shi, H. (2008). Searchable encryption revisited: Consistency properties, relation to anonymous ibe, and extensions. *Journal of Cryptology*, *21*(3), 350–391. doi:10.100700145-007-9006-6

Alliance, C. S. (2009). *Security Guidance for Critical Areas of Focus in Cloud Computing*. Retrieved from http://www.cloudsecurityalliance.org

Ballard, L., Kamara, S., & Monrose, F. (2005). Achieving efficient conjunctive keyword searches over encrypted data. *Proc. of ICICS*. 10.1007/11602897_35

Bellare, M., Boldyreva, A., & Neill, A. O. (2007). Deterministic and efficiently searchable encryption. *Proc. of CRYPTO*.

Boldyreva, A., Chenette, N., Lee, Y., & Oneill, A. (2009). Order-preserving symmetric encryption. In *Advances in Cryptology-EUROCRYPT* (pp. 224–241). Springer.

Boneh, Kushilevitz, Ostrovsky, & W. E. S. III. (2007). Public key encryption that allows pir queries. *Proc. of CRYPTO*.

Boneh, D., Crescenzo, G. D., Ostrovsky, R., & Persiano, G. (2004). Public key encryption with keyword search. *Proc. of EUROCRYPT*.

Boneh, D., & Waters, B. (2007). Conjunctive, subset, and range queries on encrypted data. *Proc. of TCC*, 535–554.

Brinkman, R. (2007). *Searching in encrypted data*. PhD thesis.

Brinkman. (2007). *Searching in encrypted data*. PhD thesis.

Cao, N., Wang, C., Li, M., Ren, K., & Lou, W. (2014). Privacy-preserving multikeyword ranked search over encrypted cloud data. *IEEE Transactions on Parallel and Distributed Systems*, *25*(1), 222–233. doi:10.1109/TPDS.2013.45

Cao, N., Wang, C., Li, M., Ren, K., & Lou, W. (2014). Privacy-preserving multikeyword ranked search over encrypted cloud data. *IEEE Transactions on Parallel and Distributed Systems*, *25*(1), 222–233. doi:10.1109/TPDS.2013.45

Cao, Wang, & Li, Ren, & Lou. (2014). Privacy-preserving multikeyword ranked search over encrypted cloud data. *IEEE Transactions on Parallel and Distributed Systems*, *25*(1), 222–233.

Cash, J., & Jutla, K., Ro3u, & Steiner. (2013). Highly-scalable searchable symmetric encryption with support for Boolean queries. *Proc. CRYPTO*, 353-373.

Chang, Y.-C., & Mitzenmacher, M. (2005). Privacy preserving keyword searches on remote encrypted data. *Proc. of ACNS*. 10.1007/11496137_30

Curtmola, R., Garay, J. A., Kamara, S., & Ostrovsky, R. (2006). Searchable symmetric encryption: improved definitions and efficient constructions. *Proc. of ACM CCS*. 10.1145/1180405.1180417

De Capitani di Vimercati, S., Foresti, S., & Samarati, P. (2012). Managing and accessing data in the cloud: Privacy risks and approaches. CRiSIS, 1 –9.

Dinh, Lee, Niyato, & Wang. (2013). A survey of mobile cloud computing: Architecture, applications, and approaches. *Wireless Commun. Mobile Comput.*, *13*(18).

Gentry, C. (2009). *A fully homomorphic encryption scheme* (Ph.D. dissertation). Stanford University.

Goh, E.-J. (2003). *Secure indexes*. Retrieved from http://eprint.iacr.org/2003/216

Goldreich, O., & Ostrovsky, R. (1996). Software protection and simulation on oblivious rams. *Journal of the Association for Computing Machinery*, *43*(3), 431–473. doi:10.1145/233551.233553

Golle, P., Staddon, J., & Waters, B. (2004). Secure conjunctive keyword search over encrypted data. *Proc. of ACNS*, 31–45. 10.1007/978-3-540-24852-1_3

Hwan, Gelogo, & Park. (2012). Next Generation Cloud Computing Issues and Solutions. *International Journal of Control and Automation, 5.*

Hwang, Y., & Lee, P. (2007). *Public key encryption with conjunctive keyword search and its extension to a multi-user system*. Pairing. doi:10.1007/978-3-540-73489-5_2

Jung, T., Mao, X., Li, X., Tang, S.-J., Gong, W., & Zhang, L. (2013). Privacy preserving data aggregation without secure channel: multivariate polynomial evaluation. *Proceedings of INFOCOM*, 2634–2642. 10.1109/INFCOM.2013.6567071

Kamara & Lauter. (2010). Cryptographic cloud storage. In *RLCPS*. Springer.

Kamara, S., & Lauter, K. (2010). Cryptographic cloud storage. In RLCPS. Springer. doi:10.1007/978-3-642-14992-4_13

Kamara, S., & Lauter, K. (2010). Cryptographic cloud storage. In RLCPS. Springer. doi:10.1007/978-3-642-14992-4_13

Katz, J., Sahai, A., & Waters, B. (2008). *Predicate encryption supporting disjunctions, polynomial equations, and inner products. Proc. of EUROCRYPT*. doi:10.1007/978-3-540-78967-3_9

Khan, Wang, Kulsoom, & Ullah. (2013). Searching Encrypted Data on Cloud. *International Journal of Computer Science Issues, 10*(6).

Khan, Wang, Kulsoom, & Ullah. (2013). Searching Encrypted Data on Cloud. *International Journal of Computer Science Issues, 10*(6).

Lewko, A., Okamoto, T., Sahai, A., Takashima, K., & Waters, B. (2010). Fully secure functional encryption: Attribute-based encryption and (hierarchical) inner product encryption. *Proc. of EUROCRYPT*. 10.1007/978-3-642-13190-5_4

Li, H., Dai, Y., Tian, L., & Yang, H. (2009). Identity-based authentication for cloud computing. In Cloud Computing. Berlin, Germany: Springer-Verlag. doi:10.1007/978-3-642-10665-1_14

Li, H., Liu, D., Dai, Y., Luan, T. H., & Shen, X. (2014). Enabling efficient multi-keyword ranked search over encrypted cloud data through blind storage. *IEEE Transactions on Emerging Topics in Computing*. doi:10.1109/TETC.2014.2371239

Li, J., Wang, Q., Wang, C., Cao, N., Ren, K., & Lou, W. (2010). Fuzzy keyword search over encrypted data in cloud computing. Proc. of IEEE INFOCOM'10 Mini-Conference. doi:10.1109/INFCOM.2010.5462196

Li, R., Xu, Z., Kang, W., Yow, K. C., & Xu, C.-Z. (2014). Efficient multikeyword ranked query over encrypted data in cloud computing. *Future Generation Computer Systems*, *30*, 179–190. doi:10.1016/j.future.2013.06.029

Liang, Cai, Huang, Shen, & Peng. (2012). An SMDP-based service model for interdomain resource allocation in mobile cloud networks. *IEEE Trans. Veh. Technol.*, *61*(5).

Liang, H., Cai, L. X., Huang, D., Shen, X., & Peng, D. (2012). An smdpbased service model for interdomain resource allocation in mobile cloud networks. *IEEE Transactions on Vehicular Technology*, *61*(5), 2222–2232. doi:10.1109/TVT.2012.2194748

Mahmoud & Shen. (2012). A cloud-based scheme for protecting source-location privacy against hotspot-locating attack in wireless sensor networks. *IEEE Trans. Parallel Distrib. Syst.*, *23*(10).

Mahmoud, M. M., & Shen, X. (2012). A cloud-based scheme for protecting source-location privacy against hotspot-locating attack in wireless sensor networks. *IEEE Transactions on Parallel and Distributed Systems*, *23*(10), 1805–1818. doi:10.1109/TPDS.2011.302

Mell & Grance. (2011). The nist definition of cloud computing (draft). *NIST Special Publication, 800*, 145.

Naveed, Prabhakaran, & Gunter. (2014). Dynamic searchable encryption via blind storage. *Proceedings - IEEE Symposium on Security and Privacy*, 639–654.

Paillier, P. (1999). Public key cryptosystems based on composite degree residuosity classes. Eurocrypt, 223–238. doi:10.1007/3-540-48910-X_16

Pang, Shen, & Krishnan. (n.d.). Privacy-preserving similarity-based text retrieval. ACM Transactions on Internet Technology, 10(1), 4.

Reddy. (2013). Techniques for Efficient Keyword Search in Cloud Computing. *International Journal of Computer Science and Information Technologies, 4*(1).

Ren, K., Wang, C., & Wang, Q. (2012). Security Challenges for the Public Cloud. *IEEE Internet Computing*, *16*(1), 69–73. doi:10.1109/MIC.2012.14

Samanthula, B. K., Elmehdwi, Y., & Jiang, W. (2014). *k-nearest neighbor classification over semantically secure encrypted relational data*. eprint arXiv:1403.5001

Shen, E., Shi, E., & Waters, B. (2009). Predicate privacy in encryption systems. *Proc. of TCC*.

Shen, Q., Liang, X., Shen, X., Lin, X., & Luo, H. (2014). Exploiting geodistributed clouds for e-health monitoring system with minimum service delay and privacy preservation. *IEEE Journal of Biomedical and Health Informatics, 18*(2), 430–439. doi:10.1109/JBHI.2013.2292829 PMID:24608048

Shen, Liang, Shen, Lin, & Luo. (2014). Exploiting geodistributed clouds for a e-health monitoring system with minimum service delay and privacy preservation. *IEEE J. Biomed. Health Inform.*, *18*(2).

Singhal, A. (2001). Modern information retrieval: A brief overview. *A Quarterly Bulletin of the Computer Society of the IEEE Technical Committee on Data Engineering*, *24*(4), 35–43.

Singhal, A. (2001). Modern information retrieval: A brief overview. *A Quarterly Bulletin of the Computer Society of the IEEE Technical Committee on Data Engineering*, *24*(4), 35–43.

Singhal, A. (2001). Modern information retrieval: A brief overview. *A Quarterly Bulletin of the Computer Society of the IEEE Technical Committee on Data Engineering*, *24*(4), 35–43.

Song, D., Wagner, D., & Perrig, A. (2000). Practical techniques for searches on encrypted data. *Proc. of S&P.*

Song, D. X., Wagner, D., & Perrig, A. (2000). Practical techniques for searches on encrypted data. In *Proceedings of S&P.* IEEE.

Song, D. X., Wagner, D., & Perrig, A. (2000). Practical techniques for searches on encrypted data. *Proceedings of S&P*, 44–55.

Stefanov, Papamanthou, & Shi. (2014). Practical dynamic searchable encryption with small leakage. *Proc. NDSS*. doi:10.1109/TPDS.2013.282

Sun, Wang, Cao, Li, Lou, Hou, & Li. (2013). Verifiable privacy-preserving multi-keyword text search in the cloud supporting similarity-based ranking. *IEEE Transactions on Parallel and Distributed Systems*. DOI: 10.1109/TPDS.2013.282

Sun, W. (2013). Privacy-preserving multi-keyword text search in the cloud supporting similarity-based ranking. Proc. 8th ACM SIGSAC Symp.Inf., Comput. Commun. Secur., 71-82.

Sun, Wang, Cao, Li, Lou, Hou, & Li. (2013). Verifiable privacy-preserving multi-keyword text search in the cloud supporting similarity-based ranking. *IEEE Transactions on Parallel and Distributed Systems*. DOI: 10.1109/TPDS.2013.282

Vaquero, L. M., Rodero-Merino, L., Caceres, J., & Lindner, M. (2009). A break in the clouds: Towards a cloud definition. *ACM SIGCOMM Comput. Commun. Rev.*, *39*(1), 50–55. doi:10.1145/1496091.1496100

Vaquero, L. M., Rodero-Merino, L., Caceres, J., & Lindner, M. (2009). A break in the clouds: Towards a cloud definition. *ACM SIGCOMM Comput. Commun. Rev.*, *39*(1), 50–55. doi:10.1145/1496091.1496100

Vaquero, L. M., Rodero-Merino, L., Caceres, J., & Lindner, M. (2009). A break in the clouds: Towards a cloud definition. *ACM SIGCOMM Comput.Commun. Rev.*, *39*(1), 50–55. doi:10.1145/1496091.1496100

Wang, C. (2010). Secure Ranked Keyword Search Over Encrypted Cloud Data. *Proc. ICDCS '10*. 10.1109/ICDCS.2010.34

Wang, C., Cao, N., Li, J., Ren, K., & Lou, W. (2010). Secure ranked keyword search over encrypted cloud data. In *Proceedings of ICDCS*. IEEE. 10.1109/ICDCS.2010.34

Wang, C., Cao, N., Li, J., Ren, K., & Lou, W. (2010). Secure ranked keyword search over encrypted cloud data. *Proc. of ICDCS'10*. 10.1109/ICDCS.2010.34

Wang, Yu, Lou, & Hou. (2014). Privacy-preserving multi-keyword fuzzy search over encrypted data in the cloud. *Proceedings - IEEE INFOCOM*.

Williams, P., Sion, R., & Carbunar, B. (2008). Building castles out of mud: practical access pattern privacy and correctness on untrusted storage. ACM CCS, 139–148. doi:10.1145/1455770.1455790

Witten, Moffat, & Bell. (1999). *Managing Gigabytes: Compressing and indexing documents and images*. Morgan Kaufmann Publishing.

Witten, Moffat, & Bell. (1999). *Managing gigabytes: Compressing and indexing documents and images*. Morgan Kaufmann Publishing.

Wong, W. K., Cheung, D. W., Kao, B., & Mamoulis, N. (2010). Secure kNN computation on encrypted databases. Proc. ACM SIGMOD Int. Conf. Manage. Data, 139-152.

Wong, W. K., Cheung, D. W., Kao, B., & Mamoulis, N. (2009). Secure knn computation on encrypted databases. *Proc. of SIGMOD*. 10.1145/1559845.1559862

Wong, W. K., Cheung, D. W.-l., Kao, B., & Mamoulis, N. (2009). Secure knn computation on encrypted databases. *Proceedings of SIGMOD International Conference on Management of Data*, 139–152. 10.1145/1559845.1559862

Yang, L., Liu, & Yang. (2014). Secure dynamic searchable symmetric encryption with constant document update cost. *Proc.GLOBECOM*.

Yang, Li, Liu, Yang, & M. (2014). Secure dynamic searchable symmetric encryption with constant document update cost. In *Proceedings of GLOBCOM*. IEEE.

Yu, J., Lu, P., Zhu, Y., Xue, G., & Li, M. (2013). Towards secure multikeyword top-k retrieval over encrypted cloud data. *IEEE Transactions on Dependable and Secure Computing*, *10*(4), 239–250. doi:10.1109/TDSC.2013.9

Yu, J., Lu, P., Zhu, Y., Xue, G., & Li, M. (2013). Towards secure multikeyword top-k retrieval over encrypted cloud data. *IEEE Transactions on Dependable and Secure Computing*, *10*(4), 239–250. doi:10.1109/TDSC.2013.9

Zhang, B., & Zhang, F. (2011). An efficient public key encryption with conjunctive-subset keywords search. *Journal of Network and Computer Applications*, *34*(1), 262–267. doi:10.1016/j.jnca.2010.07.007

KEY TERMS AND DEFINITIONS

Big Data: Large data sets that are analyzed computationally.
Cipher-Text: Encrypted form of text.
Cloud Computing: Provides shared computing resources.
Clusters: Unit of allocable hard disk space.
Encryption: Process of converting information into code.
Hierarchical Clustering: Which builds a hierarchy of clusters.
Minimum Hash Sub-tree: Root is having the minimum value in the subtree.
Symmetric Scheme: Algorithm used for cryptography.

Chapter 6
Research Analysis of Development Pipelines in Augmented and Virtual Reality Technologies

Pronay Peddiraju
VIT University, India

P. Swarnalatha
VIT University, India

ABSTRACT

The purpose of this chapter is to observe the 3D asset development and product development process for creating real-world solutions using augmented and virtual reality technologies. To do this, the authors create simulative software solutions that can be used in assisting corporations with training activities. The method involves using augmented reality (AR) and virtual reality (VR) training tools to cut costs. By applying AR and VR technologies for training purposes, a cost reduction can be observed. The application of AR and VR technologies can help in using smartphones, high performance computers, head mounted displays (HMDs), and other such technologies to provide solutions via simulative environments. By implementing a good UX (user experience), the solutions can be seen to cause improvements in training, reduce on-site training risks and cut costs rapidly. By creating 3D simulations driven by engine mechanics, the applications for AR and VR technologies are vast ranging from purely computer science oriented applications such as data and process simulations to mechanical equipment and environmental simulations. This can help users further familiarize with potential scenarios.

DOI: 10.4018/978-1-5225-3643-7.ch006

INTRODUCTION

Background of Study

Augmented and Virtual Reality allows a user to experience computer generated environments and simulations in an immersive perspective view using head mounted devices (HMDs). This project aims at shedding light at some of the development pipelines involved in creating Augmented and Virtual Reality products. By leveraging software aimed at creating these simulative environments involving 3D assets and computer-generated levels, these Augmented Reality (AR) and Virtual Reality (VR) products can be used to provide solutions to numerous real world problems. These solutions range from environment simulators to training software that can be used by corporates and individuals to reduce costs and improve efficiency. The various pipelines used in creating these products include 3D asset development, Level Design using a Game Engine, implementation of required audio and acoustics, creating the required deliverable for the target platform and the final implementation in the operating environment.

The scope for AR and VR technologies can be seen in numerous fields ranging from game development to creating training simulations that can be used in the industry to cut operating costs as well as visualize final products in real time. Industries that can benefit from these technologies however will see the requirement of a heavy initial investment depending on the complexity and the demand of the product created. Typically AR applications are less resource hungry and can be used on smartphones and have a lower initial investment while most VR applications are bulky, resource hungry and need powerful computing environments paired with precise HMDs hence making it a more expensive yet smoother experience.

This thesis aims at understanding the development pipelines involved in making AR and VR solutions and providing a development model catered to the same. From a software engineering perspective, there are numerous models that are applicable for use in the AR and VR domain but to obtain the best possible result, it is important to have a model that is specific for these types of solutions. By conducting a thorough study on the pipelines involves, we can derive some conclusions that can help us understand the requirement of resources for each and also identify the workflow in the process. The workflow can then be further simplified by implementing a parallel model while developing the various components involved in the process.

PROBLEM STATEMENT

Augmented and Virtual Reality have great scope when applied onto mobile device hardware as it can reach out to a larger user base (Jason Jong Kyu Park, 2013). But due to the complex nature of the development process and the technology itself, it can be difficult to identify a solid model that can be used to develop, maintain and re-iterate if required on the software. Engine optimizations continue to be made to allow more precise colours, details, tracking and operation (Jacob B. Madsen, 2014). This allows the development process to continue to get simpler with respect to using a game engine but there is no method that correlates the work flow between each of the pipelines. The evolution of VR and AR is extremely fast paced and is witnessing the extinction of numerous software add-ons (Dey, 2016). There is a requirement to be able to modify and re-iterate on existing products in order to ensure the scalability and longevity of the said product. There are numerous restrictions on the polygon count of the 3D models that can be successfully rendered in a VR environment due to restrictions on the hardware capabilities of hand held devices (Ahmad Hoirul Basori, 2015). This is to be identified properly and accounted for before the start of the development process to ensure that no issues with respect to support over the platform is encountered. It is important to create a smart UI (User Interaction) that is both aware of position and orientation of the user in order to provide a better user experience within the application (Hsin-Kai Wu, 2103). A common problem faced in the making of VR and AR applications is providing a good UI and UX (User Experience).

By conducting a review of the problems identified above, we can account for each and propose a model that can identify with the VR and AR development pipeline making sure that each of the pipelines involved can be handled in a way to ensure minimization of the issues listed in the above paragraph.

MOTIVATION

The motivation behind this study is to understand the development process revolving around VR and AR technologies and leveraging the tools in an efficient manner. By creating VR and AR based simulations, we are able to create a more realistic 3D perspective that the users can find a lot more comfortable to orient themselves to. By the virtue of this study, we will try to compare and contrast various methodologies for the application development process implementing VR and AR and eventually try to conceptualize a model for the same.

From a software engineering point of view, it is difficult to categorise VR and AR applications and implement a specific model for development due to its complex and flexible nature. To be able to visualize the development process and come to conclusions on the time line and nature of the project, it is important that research on the same is performed to implement a more efficient methodology or follow a model designed specifically for these applications.

By further understanding the working principles of VR and AR, we can derive a specific model for the development of the same. The model will account for all the development processes involved and hence provide a holistic view of the application development.

DEVELOPMENT PIPELINES

Although there is no specific rules in developing for VR and AR applications, the most common approach is to make use of a Game Engine paired with some 3D modelling software. The common approach revolves around the following processes:

- **Product Conceptualization:** The process of understanding requirements of the product. Analysis on its operating environment. Planning the required resources that will be involved. Implementing the required SDK for the target platform. Understanding the scope of the project and its future support.
- **Asset Development (2D or 3D):** The process of developing the required assets that will be used in the application. This involves creating 2D images, 3D object and even the generated animations, textures and other such elements that go into the game engine.
- **Audio Engineering:** The process where the required audio is created with respect to its use case and application. The audio is designed and exported in a format that can be implemented on the game engine.
- **Level Design:** The process of designing the 3D level that the user will be a part of. This involves placement of assets in the vicinity of the user and implementing the required assets at the required places within the application in a 3D space.
- **Implementing Required SDKs:** Linking the required SDK within the project to allow references to be made within the application. The SDK will provide features specific to the technology or platform that will be in use.
- **Programming Application Logic:** The process of linking all the assets and the level with logic that governs their working in the application. Usually this process goes hand-in-hand with the level design process. The application logic will define the working principles of every element in the application

and can range from anywhere between 1 simple script to several scripts that are applied on a variety of scene elements.

- **Analysis of Operating Environment:** The analysis of the operating environment to account for specific requirements that may arise for the implementation of the product in the said environment. In this phase of the development, the required changes will be made to the application in order to allow support for the target environment.
- **Testing on Target Platform:** This is a vital process in the development cycle. Upon development of the software, the testing phase identifies any issues with the application. Testing on the target platform will allow the developers to interact with application and identify and changes that are to be made in order to have the smooth functioning of the product before deployment.
- **Deployment:** The phase that involves shipping of the product to the target user base or client. The deployment phase requires the packaging of the project and ensuring an easy installation onto the target platform. It should account for any cases that may not allow a smooth installation and implement changes accordingly to make the process complete successfully.
- **Support:** The post deployment phase that ensure that any bugs or issues encountered can be dealt with by taking the required steps and making the necessary changes to the product. It allows the developers to roll out updates and fixes that maintains the smooth working of the product by increasing both the efficiency and longevity.

Each of these development processes will be discussed more elaborately in the coming sections of this document. There may be additional steps involved in development of VR and AR applications besides those listed above. For the purpose of our research, we will only be accounting for the most commonly used approach which will implement the above steps in most cases.

In this document, we will observe the steps involved in each process, the study of what tasks may be performed parallel to others so as to improve efficiency, methods that can optimize the mentioned pipeline and devise a model that can be implemented for the same. Since all the integration and use of assets is performed in the Game Engine that is implemented for development, the focus will primarily be on how to improve the process with the Game Engine in perspective.

NEED FOR DEVELOPMENT MODEL

By taking apart each of the pipelines for development, we can get a solid understanding of the steps involved in making assets and integrating them with the product. The study will apply this concept in understanding which of the pipelines can be worked upon simultaneously and which of the pipelines need to be given a greater priority.

The model shall provide an insight to the application flow and how to create a progress methodology that reduces risk, cost and other potential elements that can cause drawbacks in the development process. At the same time, the model will also provide a framework that can be implemented to improve efficiency and performance over the target platform by performing the required software planning.

To come to conclusion about the above mentioned process and derive a model that can be used, we will be implementing a VR application to understand the same. The proposed system shall be applying the use of VR on a smartphone and implement the use of 3D objects as well as audio within the application.

Introduction to Related Concepts

The proposed system is applying the use of a simple hand-held smartphone combined with a low cost software solution that may be implemented by using the said smartphone for AR and VR applications. By leveraging the Unity3D Game Engine features and combine those with those provided by the GoogleVR SDK for unity, we are able to create a variety of 3D simulative spaces that can be both interactive and immersive. The control that user has over the application is restricted to a single button interaction and the user point-of-view based visual selection.

Figure 1 is a screenshot providing a more concrete view of the GoogleVR Reticle Pointer being used in the scene to create point of view based selections.

Figure 1. VR scene view

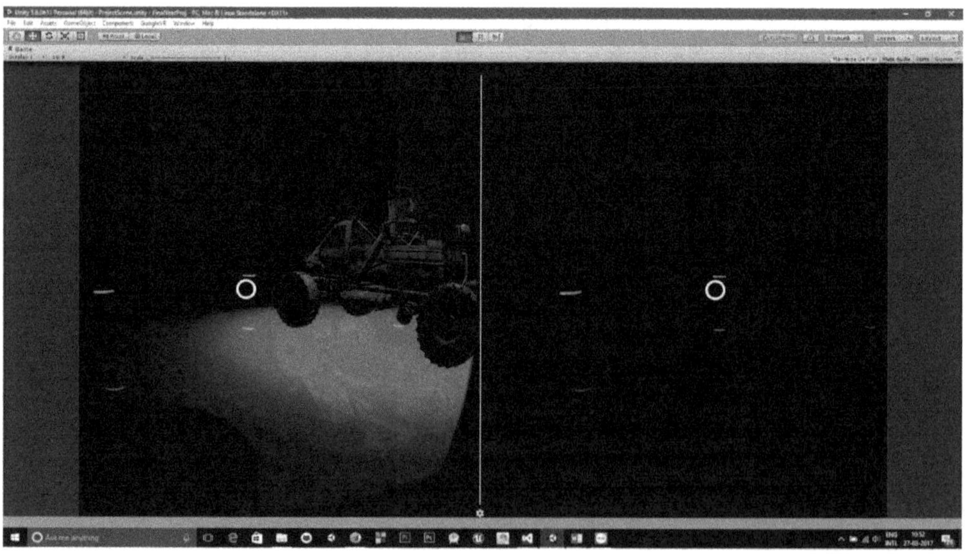

VR Camera Setup

In order to create a realistic 3D perspective view, we require the use of 2 cameras in the designed scene. As can be seen in Figure 1, the VR view provides a render using both the left eye and the right eye camera which is later merged into 1 single perspective view that provides information on both lighting and depth. This allows the user to experience a very realistic environment.

VR Reticle Pointer

The VR Reticle pointer is used to provide the user a point on the centre of the screen so that the user can identify where his point-of-view is and what objects within his view are intractable. As can be observed from Figure 1, the user reticle (white circle on the centre of the screen) will scale up and form a circle when there is an intractable object on the scene before the user. In this case we see the teleportation cylinder being highlighted to green when the user looks towards it.

VR View Port

The view port for VR implements the use of the 2 cameras as discussed. The 2 cameras provide 2 views on the view port that will create a single image when used in a VR headset (Due to the human perspective). The view port does not only account for the generated views but also provides depth, contrast and provides realistic imagery based on scene lighting.

Lighting

Lighting is among the most important elements within the level designed. If the lighting within the scene is not done properly, it could affect the look of the level and hence affect the way the objects on the scene are perceived by the user. With low quality lighting, the objects on the scene will not provide the same levels of realism which in turn creates a poor user experience within the application.

User Interface

The user interface in case of VR is slightly different compared to the regular UI implemented while developing games using a game engine. The UI for VR or AR applications must rely on world space and not create an overlay on the screen which is not ideal for all scenarios. The UI must also be placed in a way that it allows the user to have a clear view of it within the game scene. A good practice is to ensure that the UI is always facing the user irrespective of his orientation. This will allow the user to view the UI with greater clarity in world space.

LITERATURE REVIEW

From the mentioned reference documents, we are able to obtain an evolutionary view on the VR and AR development processes over the years. Some of the gaps identified from these documents are listed as follows

1. Interaction with VR and AR technologies is still restricted in most cost effective implementation cases. Due to the complexity of the product and the development process, its implementation is usually restricted if the initial investment is low. This can be countered with better planning and use of an optimized model that can allow cost cutting measure allowing for investment in additional technologies that can help improve the user interaction.
2. The use of VR and AR is still not common as the initial investment for the use of these technologies is relatively high. Again an issue that is evident due to the complexity of the development process and cost of technology required to implement the product. This cannot necessarily be countered by implementation of a development model but the model may create scope for implementing a solution that can reduce the investment making it a more cost friendly implementation.
3. There are many limitations in terms of hardware that restrict the ability of developers to create VR and AR based solutions. This requires the need for optimization and efficiency. The optimizations that are to be made should account for the operating environment and by proposing a valid model, the planning and development can be done by making note of the same.
4. VR may not always seem to be a practical solution as level complexity on the engine increases. This makes it harder for a developer to identify and modify the scenes and levels created within the game engine. By using a model that implements a proper planning phase and implementing a reference document alongside the development process, the above gap can be accounted for.
5. There are limitations with use of physics and applying forces on objects due to the orientation and conditions created by using the headset. Users encounter psychological stresses and may even experience nausea as the VR developer has the power to simulate a world before the viewer. The addition of forces in the scene or manipulating the camera in a way that affects the user's position should hence be avoided. These conditions can be implemented by compiling a valid development methodology.

6. There is still no proper solution for the problem of nausea and some other psychological effects observed by AR and VR users after using the application over long durations. Although steps can be taken to avoid this problem by implementing a valid development methodology, due to the subjective nature of this problem, sufficient data must be obtained before development of the product. This will allow the developers to account for the various results obtained on how and why these experiences are observed. Hence making a more reliable and usable software.

ARCHITECTURE IMPLEMENTED

Figure 2 is the block diagram describing the architecture of the working model used to perform analysis on various aspects of the pipeline.

Figure 2 describes the architecture of the Augmented Reality platform that will be used to analyse the steps and processes of all the pipelines involved in its development.

Given below is a brief description of the functions of each of the subsystems present in the proposed system architecture diagram:

Figure 2. Proposed system architecture.

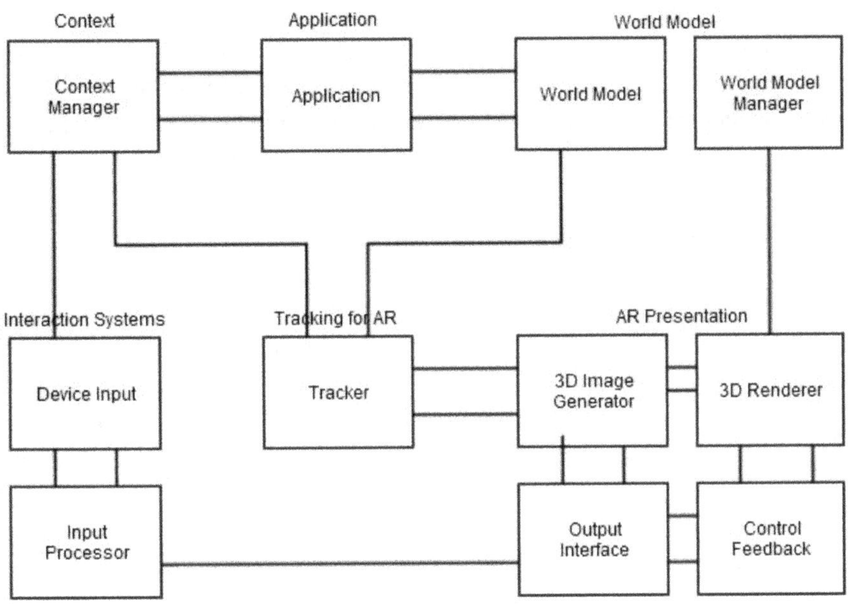

- **Application Subsystem:** The system that is the chunk of code that defines the application and its working methodologies combined with a user interface for operation.
- **Context Subsystem:** The subsystem that collects context data from various subsystems. Examples include preferences and progress data.
- **World Model Subsystem:** The subsystem that controls the use of real world elements in the 3D world space within the application. Implementation of this subsystem is the major difference between Augmented Reality and Virtual Reality based solutions.
- **Tracking Subsystem:** This subsystem used the sensors present on the device to track the user's position with respect to the 3D space. Tracking is the module that provides smooth video feed and movement flow in the application during run time.
- **Interaction Subsystem:** The subsystem that controls the user input and output. All forms of interaction between the user and the system is controlled by this subsystem.
- **Presentation Subsystem:** This subsystem is responsible for how the application is generated and presented to the user. It includes modules such as the render engine and texture repositories.

IN-DEPTH ANALYSIS

To understand the 6 subsystems in the proposed model we shall go through their respective functional diagrams. The functioning and interaction between each subsystem is integral in ensuring the smooth flow of the application and to ensure that nothing within the application is asynchronous at any given point.

Application Subsystem

To understand the application subsystem, we will refer to Figure 3 in order to observe the links and flow within the system. In this subsystem the code interacts with the various libraries and referenced game objects to provide meaning to the implementations. There may be numerous co-routines, singletons and other such entities within the application's source code for controlling a chunk of the application. An example of a singleton would be the use of a game manager code script that can only exist as a single instance within the scene designed.

The code will interact with both the Engine editor and Engine headers in order to provide functionality within the scene. Additionally the code will also work alongside the used SDKs and Plugins such as iTween (a commonly used tool for providing

Figure 3. Presentation Subsystem

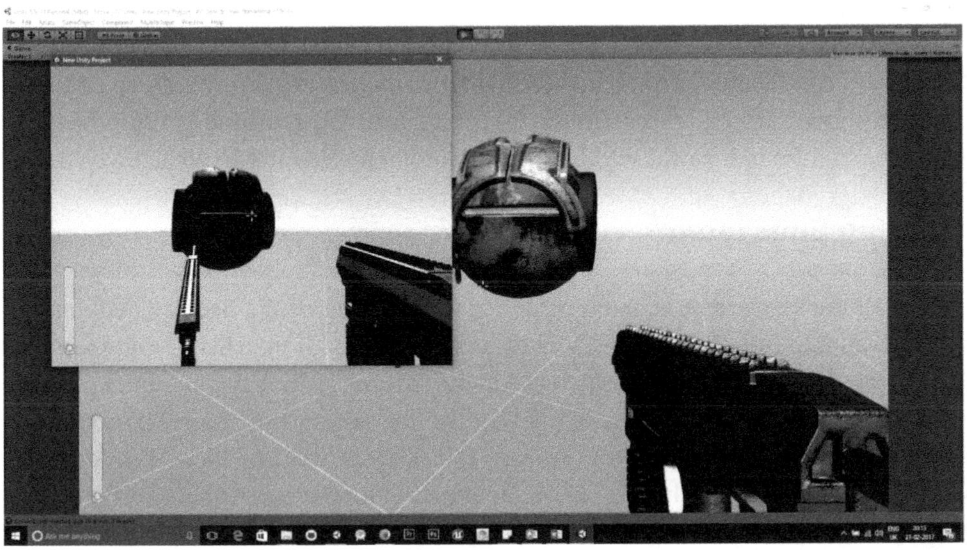

game object based transitions and animations). The references to the SDKs provide the required SDK features within the application such as the use of the VR Reticle and the dual Scene Camera setup.

Context Subsystem

The Context Subsystem communicates with all the subsystems to put together context settings and details for the application including the preferences, progress data and the in app setting that have been setup by the user. This subsystem performs a rough system analysis to understand the working principles and regulates it to a recommended setting to ensure the smooth flow of the application.

The context files interact with internal system details and read pre-sets that exist within the hardware to be able to optimize working of the application for the specific working environment at hand.

World Model Subsystem

The world model subsystem is primarily used as a manager for the level designed in the application. Ensuring the working of the level correctly and maintaining the proper perspective vision within the application is handled by the World Model Subsystem. Within this subsystem, we account for real time motion tracking accounting for the world model view with respect to the rotation, orientation and movement of the user by using the smartphone platform on a head mounted device (HMD).

By implementing the link to the Tracker module in the device, the World Model Subsystem can be able to pick up reading from the tracker to provide an accurate update to the world view and ensure a stable performance in real time as any movement performed in the real world will be immediately reflected in the virtual world.

Tracking Subsystem

The Tracking subsystem is responsible for the tracking of the user by making use of the available sensors on the HMD or device used with the HMD. For example, in the case of the VR headsets such as Oculus Rift or HTC Vive, the tracking will be performed using some specialized sensors that are part of the HMD to ensure smoothness and maintaining a real time 110 degrees field of view. While in the case of the smartphone (the case in our application for the experiment), we observe that the smartphone sensors such as the gyroscopic sensor and in some cases the availability of special hardware such as Tango by Google can be implemented for spatial tracking within the application.

This subsystem mainly interacts with sensors and ensure that the feedback received from them is accurately accounted for by the remaining subsystems.

Interaction Subsystem

This subsystem uses the available input devices to provide interaction between user and application. In the case of VR and AR the interactive elements in the scene may be in great numbers but to allow user to interact with the application we have limited hardware capabilities. This is why the interaction subsystem is used to control the variety of permutations that are performed to maximize the utility of available inputs.

This subsystem can only interact with a limited set of inputs but these inputs are manipulated by conversion to individual events. The event system hence generated is used to provide the required functionalities and interaction for the user.

Presentation Subsystem

This subsystem is complex in nature due to the tasks it is responsible for handling. It provides the application presentation for the user by loading up the required textures and by making use of a rendering engine. This requires the most computational power as it generates the level that the user interacts with. Figure 3 is a sample image of a level that is generated within the Game Engine.

Figure 3 shows the render of a scene within the engine taken from the perspectives of both clients in the scene. The clients require a generated view of their own and both are required to be generated real time. The characters on the scene and the weapons on the scene are rendered by this subsystem and provides a real-time output which in this case is synchronized via a network.

PROPOSED DESIGN

Overview

To further understand the development cycle from a software development perspective, the proposed model is to explore the following in depth to come to a conclusive analysis of the process:

The methodology implemented to create AR or VR applications involves:

1. Using the Unity Game engine to create the environment (either 3D or 2D).
2. Using a 3D asset development software such as Autodesk Maya for creating 3D models and animations that can be incorporated in the environment designed on Unity 3D.
3. Performing the required SDK integration. For the case of this project, it was to link the Unity application created with the android SDK to be able to run on any ARM based Android device (not limited to ARM).
4. Setting up the run time environment on the desired device. In the case of this project that involved getting android setup on a smartphone and making the run environment support the generated Unity game in terms of texture resolution support and memory management.
5. Finally it was integrating the application on other android devices to experience the AR environment and interact with the physical world and the AR level generated through the application.

For the purpose of this project, the methodology will incorporate the use of a level created using the Unity 3D engine. By using 3D models, audio and other such assets the scene will be created to provide a realistic user experience. To incorporate the integration of various smartphone features to improve the experience, the project will be exported to support the format of the device (Windows OS for PC, Mac OS for Apple platforms, Android for Android based smartphones, iOS for iPhones, etc). The final process will be creating a build that will run as an application on the target platform and provide an interface to share data between the operating system and the VR platform in real time to provide a rich user experience.

EXPECTED RESULT

To be able to create a Virtual Reality application that runs on an android device and provides interaction with the real world and the generated level to the user. The application will highlight the pipelines for developing Virtual Reality applications and provide a rich and immersive user experience to solve a real-world problem.

The application will use a series of technologies to provide a solution to the immersive interaction experience. These technologies include spatial tracking, GPS co-ordinate based VR tracking, and real-world perspective using the generated VR level. The result of developing this application should hence provide further clarity on how integration of VR applications with the regular operating system can be performed to provide a clear and immersive user experience. This project's scope reaches out to provide a rich user experience while using software solutions by making them user friendly, interactive, simplistic and be cheaper alternatives to existing platforms.

IMPLEMENTATION

Application Design

Mentioned below is the implementation that is being developed as part of this project. To understand the working principles of the 3D development pipeline, the below mentioned work includes details from the asset development phase to the level design and application programming phase.

VR Level Design

In Figure 4, we can see the use of a 3D asset on the scene the view of which in game can be found in the Game tab of the unity 3D software. The scene is a simple construction of a base plane and 6 assets that are used as teleportation points in the game. The teleportation points work as intractable objects that will make the user teleport to the selected location. The 3D asset was made using Autodesk 3DS Max software and shall be made intractable by implementing animations that will be trigger via the user's VR reticle. This implementation displays the level design involved to some extent in a VR application.

Figure 4. Use of a 3D asset on a VR capable level.

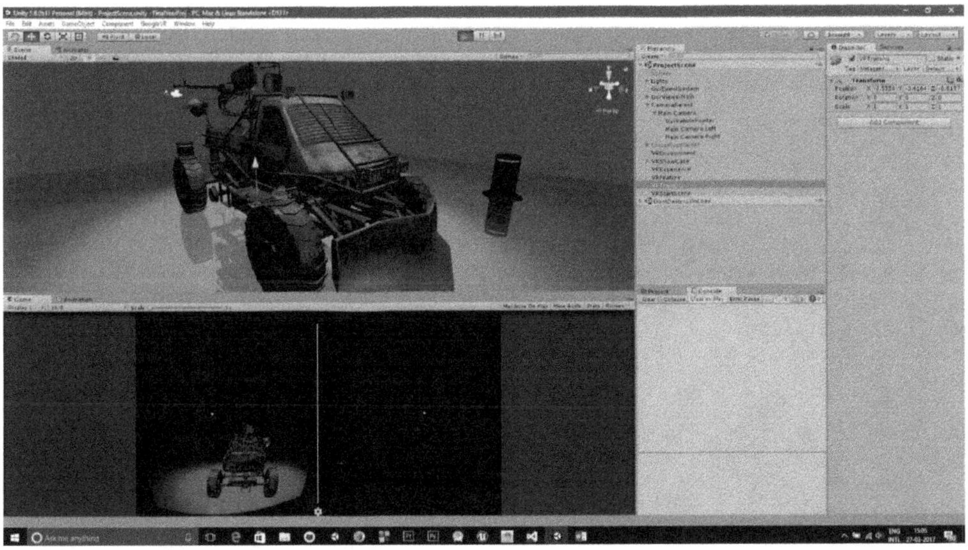

Game View VR

The game view (Figure 5) shows the use of the reticle and interactive elements that have their light turn green on hover while they remain red when they are not interacted with by the reticle. This is an example of programmed logic that is being performed on each of the interactive element that defines its working principles within the application. The level designed applies materials on the objects depending on the interaction from the user. In case of no selection performed, the object is assigned a material that provides the red glow. Once the object has been hovered over, it applies a material that provides the green glow while the reticle is still above the intractable object. After the user has hovered over the intractable element and the reticle in no longer on the object, a blue material is assigned providing a blue glow that signals that the user has seen the object and the reticle has hovered over it at some point in the application.

OPERATING ENVIRONMENT ANALYSIS

Below mentioned are all the hardware and software components used for developing the mentioned application:

Figure 5. Game View and reticle pointer for the generated VR level

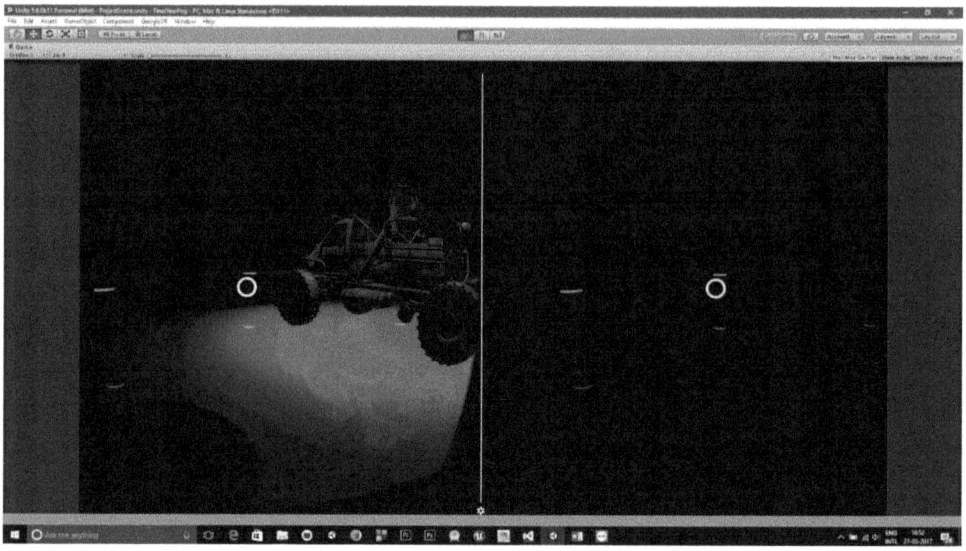

Hardware

- **Android Device:** Any android device that adheres to the minimum specification requirements in order to run applications developed on the Unity 3D engine.
- **Augmented Reality Headset:**The headset will carry your smartphone with a pair of lenses that provide an AR experience. Examples of such headsets include GearVR, VR Box and Google Cardboard.
- **Raspberry Pi (To provide a controller for the user):**If the application requires some form of user interaction, a raspberry pi can be used to provide the interaction with user by implementing IR sensors.

Software

- **Unity 3D Software:** Unity 3D is a Game Engine that can be used to develop applications that incorporate either 2D or 3D world space. This engine will be used for the purpose of this project to develop an android based scene that can provide an AR experience while using it in Google Cardboard.

- **Autodesk Maya:** Autodesk Maya is a 3D modelling software provided by Autodesk. It will be used in this project to provide all the 3D assets and animations that will be incorporated in the projects.
- **Android Studio:** Used to provide the integration of the C# based application code with the java based android platform by making use of the intents provided by android.

ACKNOWLEDGMENT

The author would like to thank the School of Computer Science and Engineering, VIT University and special thanks to Dean SCOPE, for his kind guidance and support along with our guide Dr. Prabu S and Dr. Swarnalatha P, without whom it would not have been possible to complete this mammoth task. Also, I would like to thank the anonymous reviewers and the editor-in-chief of the IGI global publishers for their valuable guidance which has improved the quality and presentation of the chapter.

REFERENCES

Basori, Afif, Almazyad, Abujabal, Rehman, & Alkawaz. (2015). Fast Markerless Tracking for Augmented Reality in Planar Environment. *3D Research, 6*(4), 1-11.

Dey, A. (2016). A Systematic Review of Usability Studies in Augmented Reality between 2005 and 2014. *Mixed and Augmented Reality (ISMAR-Adjunct),* 2016 *IEEE International Symposium on.*

Jason, J. K. P., Park, Y., & Mahlke, S. (2013). Efficient execution of augmented reality applications on mobile programmable accelerators. *Field-Programmable Technology (FPT) 2013 International Conference on*, 176-183.

Kristensen, B. B., May, D., & Nowack, P. (2011). Collaboration and Modeling in Ambient Systems: Vision Concepts and Experiments. *System Sciences (HICSS) 2011 44th Hawaii International Conference on*, 1-6.

MacWilliams, Reicher, Klinker, & Bruegge. (2014). Design Patterns for Augmented Reality Systems. *CEUR Workshop Proceedings*, 19.

Madsen, J. B., & Stenholt, R. (2014). How wrong can you be: Perception of static orientation errors in mixed reality. *3D User Interfaces (3DUI) 2014 IEEE Symposium on*, 83-90. 10.1109/3DUI.2014.6798847

Teather & Stuerzlinger. (2016). SIVARG: Spatial Interaction in Virtual/Augmented Reality and Games. Proceedings of the 2016. doi:10.1109/ISMAR-Adjunct.2016.0036

Wu, H.-K., Lee, S. W.-Y., Chang, H.-Y., & Liang, J.-C. (2013, March). Current status, opportunities and challenges of augmented reality in education. *Computers & Education*, *62*(C), 41–49. doi:10.1016/j.compedu.2012.10.024

Chapter 7
Iterative MapReduce:
i-MapReduce on Medical Dataset Using Hadoop

Utkarsh Srivastava
VIT University, India

Ramanathan L.
VIT University, India

ABSTRACT

Diabetes Mellitus has turned into a noteworthy general wellbeing issue in India. Most recent measurements on diabetes uncover that 63 million individuals in India are experiencing diabetes, and this figure is probably going to go up to 80 million by 2025. Given the rise of big data as a socio-technical phenomenon, there are various complications in analyzing big data and its related data handling issues. This chapter examines Hadoop, an open source structure that permits the disseminated handling for huge datasets on group of PCs and thus finally produces better results with the deployment of Iterative MapReduce. The goal of this chapter is to dissect and extricate the enhanced performance of data analysis in distributed environment. Iterative MapReduce (i-MapReduce) plays a major role in optimizing the analytics performance. Implementation is done on Cloudera Hadoop introduced on top of Hortonworks Data Platform (HDP) Sandbox.

INTRODUCTION

We live in the age of data. It's not easy to measure the total volume of data. IDC estimate puts the size of the "digital universe" at 4.4 zettabytes in 2013 and is forecasting a tenfold growth by 2020 to 44 zettabytes. Such enormous volumes of data suffer from various issues like storage capabilities and synchronization

DOI: 10.4018/978-1-5225-3643-7.ch007

problems since they are stored at different places depending upon the vicinity to data servers. Given the rise of Big Data as a socio-technical phenomenon there are various complications in analyzing Bigdata and its related data handling issues. In such a case Iterative MapReduce comes in really handy. The term 'Enormous Data' alludes to the monstrous volumes of both organized and unstructured information which can't be directly utilized with conventional database administration frameworks.

With the quick increment in the diabetic patients in India and number of determinants for the diabetes, the information will become tremendous and turns out to be Big Data which couldn't be handled by traditional DBMS. Here we discuss about Hadoop, an open source structure that permits the disseminated handling for huge datasets on group of PCs and thus finally produces better results with the deployment of Iterative MapReduce. The main goal is to dissect and extricate the enhanced performance of Data Analysis in distributed environment. Iterative MapReduce (i-MapReduce) plays a major role in optimizing the analytics performance. Implementation is done on Cloudera Hadoop introduced on top of Hortonworks Data Platform (HDP) Sandbox. Hortonworks Hadoop is used for the extraction of useful data patterns in light of the inquiry identified with different determinants of diabetes dataset obtained from IBM quest. Iterative MapReduce algorithm for sequential pattern mining utilizes a distributed computing environment. It consists of two processes a mapping process and reducing process which is further utilized in two separate phases namely Scanning phase and Mining phase. During the scanning phase high performance is gained by distributing the task of finding elements over different mapper tasks which can be run parallel on multiple machines with a distributed database or file system. In the mining phase the mapper task creates a lexical sequence tree for finding patterns and to improve efficiency of limited depth dfs which is run on the tree and the reducer task finds the support value for the patterns and thus in turn finds the useful patterns. So, to avoid the problem of the serialized processing, we opt for parallel processing in Big Data environment. This parallel processing not only reduces the computation time but also optimizes the resource utilization.

This is a brief idea about the chapter and we will see its implementation and impacts in further sections.

BACKGROUND

Big Data is also like normal data but with an enormous size. This is a term is generally used to describe a collection of data that is very huge in size and still growing exponentially with time. In short, such a large collection of data which is difficult to handle via traditional databases and other management tools is called 'BigData'. Generic features of BigData are:

- Volume
- Velocity
- Variability
- Veracity

The main objective of this analysis is to find interesting patterns on the basis of conditional dependence on given attributes of a dataset. With such an increased rate of data generation it becomes very difficult to analyze the patterns in the dataset. Also the situation becomes more critical when we have sequential patterns in the dataset i.e. the order of dependency matters. Such behavior of data is very usual in day to day actions such as customer shopping behavior, medical symptoms leading to a future patient disease, financial stock market data predictions etc. Pattern mining of BigData using Hadoop faces a lot of issues in terms of data storage, data shuffling, data scanning, data processing units etc.

Sequential pattern mining is one of the most important data mining technique used in various application technologies in modern world. Some examples are Gene Analysis, Intrusion detection of System attack and Customer Behaviour Prediction. The centralized logic behind sequential pattern mining is to find frequent sequences within a transactional or operational database. The formal definition can be detailed as follows. Definition 1: Let D be a sequence database, and I = {y1, · · ·, ym} be a set of m different items. S = {d1, · · ·, di} is a sequence consisting of an ordered list

Figure 1. Implications of BigData

of itemsets. An itemset di is a subset of items ⊆ I. A sequence Dr = {r1, · · ·, rn} is a subsequence of sequence Dt = {t1, · · ·, tm} where 1 ≤ i1 < · · · < in ≤ m such that r1 ⊆ ti1, r2 ⊆ ti2, · · ·, rn ⊆ tin. Sequential pattern mining is generally used to find all the sets of sequential patterns whose occurrence frequencies ≥ minimum support ∗ |D|. Here minimum support is the support threshold value. Some of the patterns which will be of great help in this regard are Apriori-based, Pattern growth-based and Projection-based algorithms.

Pattern mining algorithms are mainly of two types:

- Join based algorithm
- Tree based algorithm

Join based algorithm includes basic Apriori algorithm which is executed in recursive level-wise manner. It includes three major steps that are repeatedly executed again and again for different values of k (the length of the pattern). First step involves the generation of candidate patterns followed by the pruning of patterns and finally the validation of patterns against the given dataset. Apriori algorithm finds the enum tree by making use of bfs (breadth first search). It generates the candidate key using join-based approach. Observing the execution of Apriori, we can see that enum tree is not explicitly used but the tree after scanning phase is constructed on the basis of dataset prefixes. Other major algorithms in this field are TreeProjection and FP-growths. Both of these algorithms make use of hierarchical relations between the given dataset and the useful patterns between them. It reduces the number of data scans and thus having lesser resource requirements.

Tree based algorithm includes DHP (Direct Hashing and Pruning) which was proposed as an improvisation for Apriori algorithm. It is built on traditional logic only but with two major improvisations. The first improvisation is to prune the itemsets in each iteration, and the second improvisation is to trim the transactions so that support-counting becomes more efficient. After first scan of the dataset it maintains a DHT (Distributed Hash Table) and uses it for all further calculations. It is this DHT which is used for making a BitMap representation table for determining the presence or absence of an item set in the dataset. For presence of an item in the dataset its normalized frequency count is taken into consideration with all conditional attributes.

IBM's Watson Analytics, introduced initially for the finance industry, attempts to provide a simple analysis system based won IBM's sophisticated Cogno's processing capabilities. It uses NLP to give predictive decisions based on the input data. It was initiated with an objective to help business processes to access data remotely for coming up with business related decisions. All types of businesses generate data these days. It is done by means of websites, social media, shopping patterns of customers, user experience, etc. This reflects the need for companies to strategize on various aspects of storage and mining of useful information for the available data. This is considerably more challenging than just locating, identifying, understanding, and citing data. In order to take decisions based on the available data, the systems require being fully fledged automated which can store data and schedule tasks at regular time intervals to phase out results from time-time.

MAIN FOCUS OF THE CHAPTER

The main agenda of this chapter is to understand about traditional pattern mining algorithms and inculcating improvisations in the same for BigData processing in a distributed environment. We will see how data is stored in master and slave nodes and how they can be synchronized to operate interchangeably for improved results. This methodology will not only reduce the computation time but will also improve resource utilization. The best solution to this situation is using parallel computing along with MapReduce algorithms in a distributed environment. It involves a lot of parameters such as support count, confidence, discrete attributes, continuous attributes, leveling tags etc. On the basis of all these attributes, patterns can be classified as:

- **Negative Frequent Patterns**

One of the major challenges in pattern mining in the segregation of useful patterns from the others. It is this reason because of which different attributes are taken into consideration to come up with proper segregation. Some of them are support, confidence, normalization factor etc. Suppose an item has some frequency of occurrence in a dataset but this does not gives us a real picture of the item. There are many conditional parameters which determines the occurrence of the item. So we can say that absolute frequency of an item has no meaning in pattern mining. It needs to be conditional considering all the biases and interrelated factors. Various pattern mining approaches focus on frequency normalization and they get the correct picture of patterns in a dataset. In case of negative association, the general observation is that the occurrence of one item with greater support means the absence of other item in the dataset.

- **Constrained Frequent Patterns**

Most of the times we need to filter the patterns according to user requirements and make an appropriate choice of the distinguishing parameters. For example a client may want the presence of one specific itemset in the mining dataset but defines some differentiating factors which determine the conditional rule base for that particular mining pattern. We can in fact directly push all the constraints in the mining process. It has several advantages as mining can be performed at much lower support levels as compared to other methods. Thus we can say that constrained pattern mining involves deployment of conditional patterns with normalized frequency values.

- **Compressed Frequent Patterns**

One of major issues in pattern mining is the volume and redundancy of patterns in a given dataset. All traditional algorithms need to make multiple data scans for finding interesting patterns. This increases the operational and system requirements. Most of the times a subset of a frequent pattern set is also frequent. Also to make the process easy, most algorithms make use of bitmap representations of the useful patterns and simultaneously pruning the remaining patterns. Therefore it solves the problem of multiple data scans and also reduces the computational requirements.

The main objective of the proposed system is to make this process hassle free for the users i.e. they will not have to maintain clusters and spend huge amounts for the hardware stuff to mine information as well as schedule tasks responsible for Big Data analysis. This system is built upon the Hadoop ecosystem. It will be a mix of standalone and hosted service for the clients and will greatly reduce their workload. It aims at finding all interesting patterns in the dataset based on the user input and thus helps in categorizing the useful patterns from the others. It aims at reducing the hardware requirements for the computation and produce resource optimized results. It will greatly boost the development of any product or services in the market industry with reference to:

- Investors
- Business Analyst
- E-commerce companies
- Advertising agencies
- Wholesaler

If the transactional database has huge amounts of frequent patterns then it becomes difficult to load complete data in the main memory. So to overcome this problem various parallel mining techniques can be used which parallely mines the frequent patterns and gives the desired output. Now to solve this problem we can make use of Sequential Pattern Mining on the Cloud which makes use of cloud framework. It deploys bitmap representation to find the mining patterns. Traditional approaches operates in centralized fashion which has restricted computation power and efficiency. Sequential Pattern Mining on the Cloud provides a good solution for this problem. It makes use of recursive MapReduce techniques with two major phases. Firstly it scans the dataset and then mines the dataset. During scanning phase it scans the data and makes a distributed hash table out of it. Now in the mining phase it makes use of iterative MapReduce processes which forms simultaneous Mappers and executes the process concurrently. It deploys a distributed version of sequential pattern mining which utilizes the divide and conquer strategy. Our algorithm utilizes MapReduce computing algorithm for sequential pattern mining in a distributed environment. It consists of two processes a mapping process and reducing process which is further utilized in two separate phases namely Scanning phase and Mining phase.

In the scanning phase high performance is gained by distributing the task of finding elements over different mapper tasks which can be run parallel on multiple machines with a distributed database or file system. Reducer tasks are then run to find the frequency of different patterns and infrequent patterns we subsequently removed from the records. The result can be stored to be passed on to the mining phase.

In the mining phase the main aim is to find all possible patterns in the dataset and to prune away non-sequential and non-useful patterns to filter out the important information from the dataset. We try to achieve this using a lexical sequence tree which can generate all possible patterns existing in the dataset and then this tree can be pruned away to find out useful or sequential patterns.

So now comes a very obvious question, how is iterative MapReduce better than traditional algorithms in distributed environment. Here are some of the comparing parameters:

- **Memory Scalability:** This method greatly reduces the memory requirements because of Master-Slave architecture. Here the dynamic nature of data is also taken into consideration which helps is parallel processing. The required memory is also reduced in a way that all the unnecessary patterns are continuously being pruned off.
- **Work Partitioning:** Most of the data processing jobs are divided in two phases namely the Scanning phase and the Mining phase. Both these phases have a Mapper task and a Reducer task. Now the execution of mapper phase of the master will not hamper the execution of mapper phase of the slave nodes and vice-versa. This will eliminate the mutual dependency of various MapReduce tasks.

- **Scaling**: Setting up Hadoop is a one-time investment and after that it can be easily scaled up with new cluster nodes. It can be deployed to economical commodity servers, permitting the groups of nodes to develop as required.
- **Data Restrictions**: Hadoop can handle unstructured data without the need to process it before use. There is no need to massage the data before it can be used. So, MapReduce is preferred slightly more here.
- **Language**: The language behind MapReduce's control component is basically Java. It also makes use of internal Pig and Python which will make the process of management and deployment easier.

ISSUES, CONTROVERSIES, AND PROBLEM

With such an increasing rate of Data generation it is becoming very difficult to store all the data on a single device and compute it locally. This is the reason that Distributed computing comes into picture. The major issue with traditional pattern mining algorithms is that it needs to scan the same dataset again and again thus increasing the computational requirements. Also there is no discrete method of segregating the useful patterns from the uninteresting patterns. Most of these algorithms make use of local computation which has the constraint of storing data locally and having repeated data scans. Another application lies in the field of genome sequencing. Genome sequencing decodes the order of DNA nucleotides, or bases, in a genome—the order of A's, C's, G's, and T's that make up an organism's DNA. The human genome is constituted of over 3 billion of these genetic letters. Samples are collected from the patients suffering from genetic disorders, genome sequencing is carried out to determine the appropriate genetic abnormality in patients. These genome sequencing samples generate terabytes of data which need to analyze in real time to provide accurate results to decipher and verify with existing sequences. There are many more places of applications of data analysis in the domain of Big Data. There are many such scenarios where we require instant results from large datasets but due to various constraints we are not able to get results on time. These may be due lack of processing power, lack of skillset etc. Thus Iterative MapReduce(i-MapReduce) comes to our rescue and reduces the dependancy on Hardware modules and clusters.

So to overcome all these issues we can make use of distributed version of sequential pattern mining which utilizes the divide and conquer strategy. The main focus of the algorithm is to utilize MapReduce computing algorithm for sequential pattern mining in a distributed environment. It consists of two processes a mapping process and reducing process which is further utilized in two separate phases namely Scanning phase and Mining phase.

SOLUTIONS AND RECOMMENDATIONS

The architecture comprises of the following elements Data Storage, Data parser, Data sampler, Data analysis and finally data visualization layer.

The sampling layer is responsible for collecting random pieces of data in order to perform a few initial preprocessing tasks. It performs the task of selecting a subset from the original set of all measurements. Since unstructured data is fed into this layer so the task of structuring it is performed by the sampling layer. This is done by means of finding the delimiter of the input data and finding useful summarizations of the sampled data to get an overview of the data to run optimized algorithms on the same. The types of sampling ranges from simple random sampling which in which any particular data has equal probability of being sampled to custom sampling where the user provided how much percentage of data is to be sampled from the input data.

The data parsing layer supports extraction of various attributes of any type of data. This layer supports in various formats like CSV,XML and JSON since most of the data meant for analysis is available in any one of these formats. This layer also supports extraction of specific attributes of the data. This is again based on custom queries as provided by the user for any of the three input types.

The core analysis layer is subdivided into various sublayers. These sublayers are named aggeration, filtering, classification and data organisation layers respectively. The data enters into this layer after being sampled and parsed and hence is ready for the type of analysis which has been specified by the user. The objective of the filtering layer is to extract certain chunks of data on the basis of some applied function or randomly extracting chunks which vary in size and dimension. Major functionality in this layer is finding all distinct values present in a string of parameters so that all repetitions are filtered out thus reducing redundancy of patterns.

The objective of the visualization layer is that it helps in getting a better understanding of the data. This layer makes the picture clearer to a general user. After running a specified algorithm on the input data set, an output file will be generated which will be fed into the software named Tableau along with the cluster details and port number. This will generate a graphical representation of our output data for improved understanding.

It has two main phases:

1. Scanning Phase

During the scanning phase high performance is gained by distributing the task of finding elements over different mapper tasks which can be run parallel on multiple machines with a distributed database or file system. Reducer tasks are then run to find the frequency of different patterns and infrequent patterns we subsequently removed from the records. The result can be stored to be passed on to the mining phase. The major task in this phase is to create the lexical sequence tree. Each common node or its associated subtree represents a sequence in the data set. We gain scalability in this phase using by splitting the data set into parts and running a mapper on each independently to offer better execution time on distributed systems. The mapper runs a depth search on the tree mine the patterns but the search is a depth limited search to reduce the load on the mapper program. This removes a bottleneck in the LST construction and mining and gives more balance to each mapper phase. The mapper phase creates an intermediate output which can be processed by the reducer processed to prune away non-sequential patterns. The pattern is stored in the format<pattern, bit-AND -result> for the reducer side to process. The mapper phase outputs the patterns along with the node depth and a threshold value is assumed for the node depth any node which has support count lower than the threshold is not processed and its extensions are also not processed.

Pseudo Code for Scanning Phase (Mapper)

Client Input: p (partitioner)
Step 1: Variable p = to take input data;
Step 2: Variable data = p.value; // < Custid, Tid, Item >
Step 3: For loop each data in p **do**
Step 4: Output < data.Item, (data.Custid, data.Tid) >;
Step 5: End for loop

Pseudo Code for Scanning Phase (Reducer)

Mapper Input <k, v>
sup, threshold
Step 1: Variable i = k; // storing input
Step 2: Variable sup = s; // used to store support
Step 3: For loop each value in v **do**
Step 4: Sup = find the support count of each item (i)
Step 5: End for loop
Step 6: For loop each < i, s > in sup **do**
Step 7: If s(i).support >= sup **then**
Step 8: Variable bit = create the bitmap of each item (i)
Step 9: Put < i, (data, bit) > to DHT
Step 10: End if
Step 11: End for loop

2. Mining Phase

In the mining phase the mapper task creates a lexical sequence tree for finding patterns and to improve efficiency of limited depth dfs which is run on the tree and the reducer task finds the support value for the patterns and finds the useful patterns. These phases are further elaborated. In the scanning phase the data is loaded on to the machine for scanning and mining. To avoid loss of data during loading each mapper task and reducer task each of these tasks reads pre-partitioned part of the data based on a memory chunk or no of lines. The mapper task converts the data Tuples to key value pairs for faster processing. The tuples are grouped based on itemID and the transaction IDs are stored in the value part. For example, item,(tid,tid) The reducer task counts the support value for the items and patterns which can be used to remove infrequent patterns. For handling data between mappers and reducers the mapper and reducer with identical keys handle the same set of data. A threshold would be identified to remove the infrequent patterns. To remove infrequent patterns any pattern having confidence value less than the threshold would be removed and the reducer would store the data in a distributed hash table or a distributed file system would be accessible to the next MapReduce job in the mining phase.

Pseudo Code for Mining Phase (Mapper)

Input: X, depth (d), maximum depth of sub-tree
Step 1: Variable LST; // Lexical Sequence Tree on loc machine
Step 2: LST = generate a sub-tree with root node is X and depth d
Step 3: **For loop** each node in LST **do**
Step 4: Variable bit-A_r = perform bit-AND operation
Step 5: **If** (bit-A_r > 0)**then**
Step 6: Output < node.pattern, (Cid, bit-A_r) >;
Step 7: **End if**
Step 8: **End for**

Pseudo Code for Mining Phase (Reducer)

Mining Input: a set of < pattern, (Custid, bit-A_r) > pairs
min_sup, minimum support threshold
Step 1: **For loop** each pattern **do**
Step 2: Freq = store net support count of this pattern;
Step 3: **If** freq >= min_sup **then**
Step 4: Output pattern < pattern, freq >;
Step 5: **If** this pattern is a leaf node in the current sub-extension-tree **then**

Step 6: Output pattern to the extensible set;
Step 7: **End if loop**
Step 8: **End if loop**
Step 9: **End for loop**

Here the main aim is to find all possible patterns in the dataset and to prune away non-sequential and non-useful patterns to filter out the important information from the dataset. We try to achieve this using a lexical sequence tree which can generate all possible patterns existing in the dataset and then this tree can be pruned away to find out useful or sequential patterns. The task of the reducer phase is to merge the outputs from the mapper phase and reducer phase prunes away the patterns with low support count. The reducer phase extracts the depths and support count from the BitANDresult to gain info from the preprocessed data form the mapper phase a reducer phase works on the output of a unique mapper in parallel on a distributed environment constructed on mapper processes running in parallel. We can conclude from this behavior that larger datasets can be handled better by synchronizing master and slave nodes. The better option is to always trigger the master node with recursive slave node callouts.

Various parameters have been considered for performance evaluation in Pattern Mining on various datasets. Some of them are:

- Memory scalability
- Work partitioning
- Load balancing.

Evaluation of performance of Iterative-MapReduce on diabetes dataset using Hadoop with 1 master node and 2 slave nodes gave positive results. It took reduced memory rounds and computed the patterns very efficiently. On comparing with the existing methods this is much more scalable and accurate. Generic implementation of i-MapReduce on Hadoop 1.0.3 and jdk-7u3 in a virtual cloud environment consisting of 3 clusters gave a memory round 225ms. First cluster serves as both master and slave node and the remaining 2 clusters serve as slave nodes. All these evaluations which are carried out on machines with 2.64GHz Intel Xeon CPU, 8GB main memory, and 1GB network bandwidth produce similar outputs. It involves the process of feeding the (key, value) pair to the mapper of the master node and then the slave node keeps on repeating the iterative process till we achieve the confidence of 85% and accuracy greater than 80%. With each iterative process the slave node uses the output of first tool runner class and feeds as input for the next mapper task. On investigating the results, it can be clearly seen that larger the number of Map Reduce

Figure 2. Architecture diagram

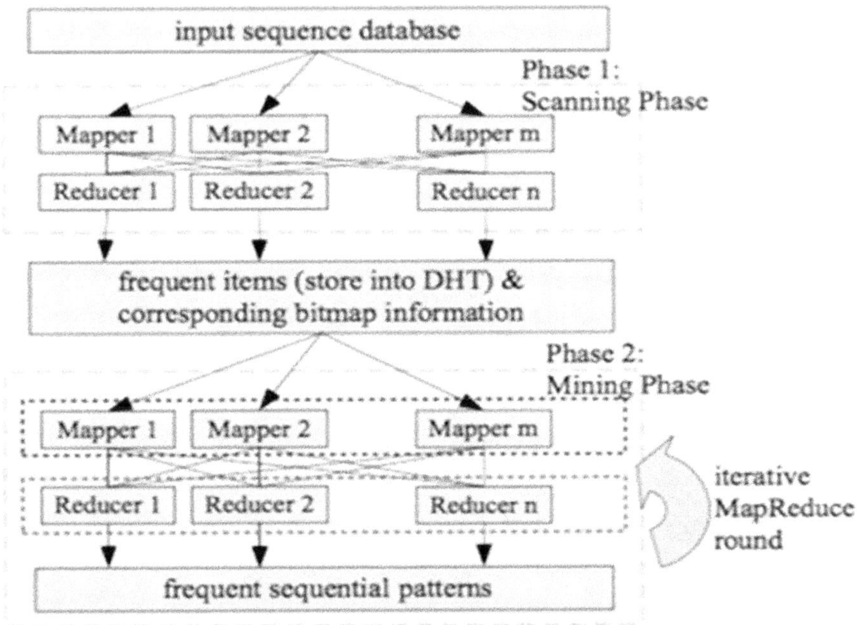

rounds, the better the Sequence tree of Mapper and thus this will help in reducing the initialization costs and making enhanced and smooth dataflow. Results have also shown that generally finding larger patterns will take more time, so Iterative MapReduce focuses on finding small patterns and remembers them so that they can be further be used to propagate on other larger patterns and this process in continued until we get the threshold accuracy and confidence value specified by the user thus eliminating all unnecessary patterns.

FUTURE RESEARCH DIRECTIONS

For future work, further investigation needs to be done on improving the computation time of the MapReduce algorithms. Clusters can be setup on parallel GPUs and then observations can be made as to how the algorithm reacts to a highly parallel environment with real time influx of data and better memory management. Also intrinsic pattern pruning is one area which has a great scope of future research. The cluster nodes need to be made smart enough to automatically prune the uninteresting patterns from the dataset.

Figure 3. Metric (Time, Space) comparison

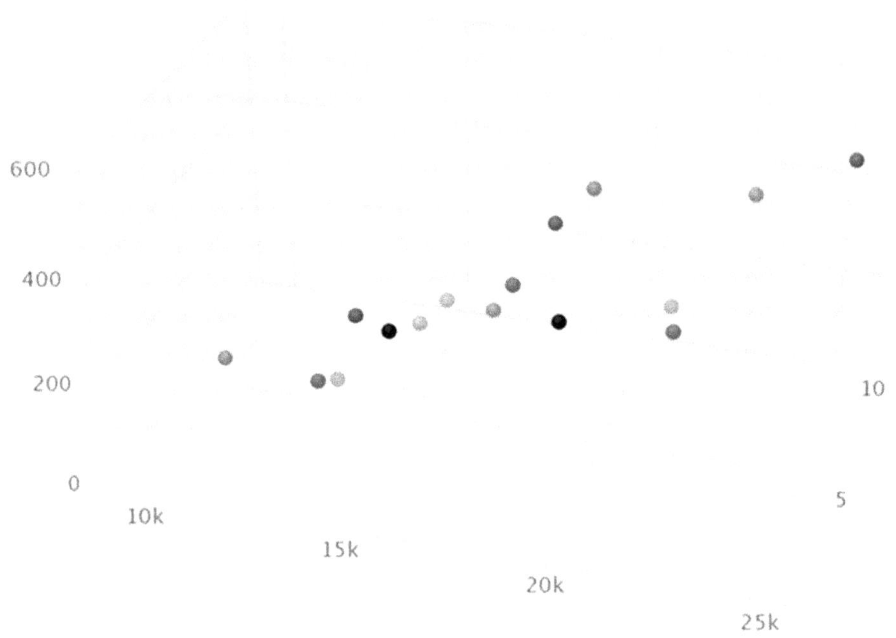

Figure 4. Diabetic dependency on genes

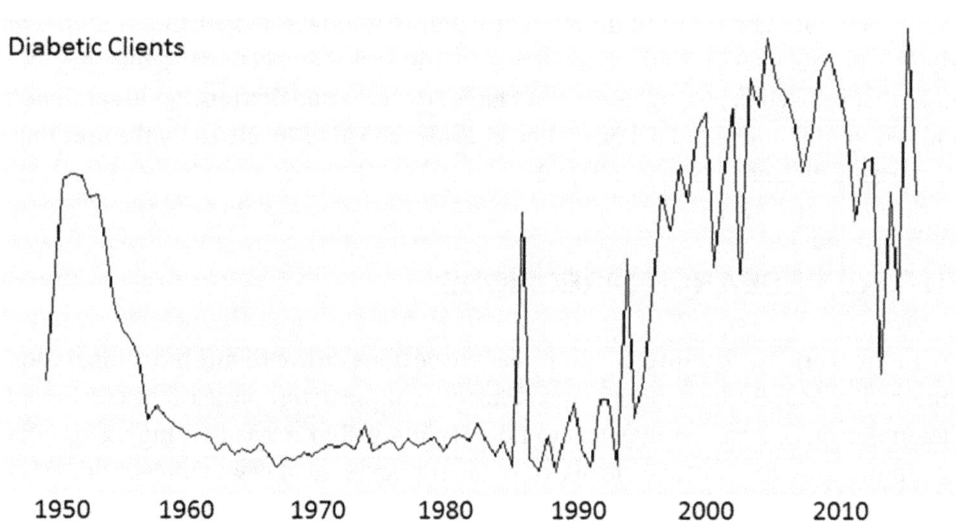

CONCLUSION

In this chapter we have discussed an optimized version of a traditional pattern mining algorithm on a distributed setup using Hadoop framework and Iterative MapReduce algorithms. A stepwise system to find and filter relevant patterns from the data set using Iterative MapReduce is the major point discussion in the chapter. The capacity to handle large data sets varies depending upon the server's capacity as well as the amount of commodity servers available. By the usage of commodity server's the data processing has been scaled out. Thus we can say that MapReduce algorithms play a very pivotal role in pattern mining. The distributed environment of the node clusters helps in smooth flow of data accompanied by efficient computation and optimized resource utilization.

REFERENCES

Dean, J., & Ghemawat, S. (2008). MapReduce: Simplified data processing on large clusters. *Communications of the ACM*, *51*(1), 107–113. doi:10.1145/1327452.1327492

Demirkan, H., & Delen, D. (2013). Leveraging the capabilities of service-oriented decision support systems: Putting analytics and big data in cloud. *Decision Support Systems*, *55*(1), 412–421. doi:10.1016/j.dss.2012.05.048

Kumar, S. (2004, January 01). MEGA3: Integrated software for Molecular Evolutionary Genetics Analysis and sequence alignment. *Briefings in Bioinformatics*, *5*(2), 150–163. doi:10.1093/bib/5.2.150 PMID:15260895

Labrinidis, A., & Jagadish, H. V. (2012). Challenges and opportunities with big data. *Proceedings of the VLDB Endowment International Conference on Very Large Data Bases*, *5*(12), 2032–2033. doi:10.14778/2367502.2367572

Reges, S., & Stepp, M. (2014). *Building Java Programs*. Pearson.

Ristoski, P., Mencía, E. L., & Paulheim, H. (2014, May). A hybrid multi-strategy recommender system using linked open data. In Semantic Web Evaluation Challenge (pp. 150-156). Springer International Publishing. doi:10.1007/978-3-319-12024-9_19

Sankaranarayanan, S., & Perumal, T. P. (2014). Diabetic Prognosis through Data Mining Methods and Techniques," *Intelligent Computing Applications (ICICA), 2014 International Conference on*, 162-166.

Taylor, R. C. (2010). An overview of the Hadoop/MapReduce/HBase framework and its current applications in bioinformatics. *BMC Bioinformatics*, 11. PMID:21210976

Tukey, J. W. (1977). *Exploratory data analysis*. Academic Press.

White. (2012). *Hadoop the definitive guide*. O'Reilly Media, Inc.

Chapter 8
Big Data for Satellite Image Processing:
Analytics, Tools, Modeling, and Challenges

Remya S.
VIT University, India

Ramasubbareddy Somula
VIT University, India

Sravani Nalluri
VIT University, India

Vaishali R.
VIT University, India

Sasikala R.
VIT University, India

ABSTRACT

This chapter presents an introduction to the basics in big data including architecture, modeling, and the tools used. Big data is a term that is used for serving the high volume of data that can be used as an alternative to RDBMS and the other analytical technologies such as OLAP. For every application there exist databases that contain the essential information. But the sizes of the databases vary in different applications and we need to store, extract, and modify these databases. In order to make it useful, we have to deal with it efficiently. This is the place that big data plays an important role. Big data exceeds the processing and the overall capacity of other traditional databases. In this chapter, the basic architecture, tools, modeling, and challenges are presented in each section.

DOI: 10.4018/978-1-5225-3643-7.ch008

1. INTRODUCTION

Day by day, we see the data is rapidly increasing in many forms. We have some traditional data processing software to process small quantity of data. But as trillions of bytes of information is being processed per second, the traditional software techniques fail in processing this data. We need to re-think of a solution which can process this data. Now Big Data gives us a solution. Big Data is a term used for creating, capturing, communicating, aggregating, storing and analyzing large amounts of data. Many attempts encountered to quantify the growth rate in the volume of data is called as Information Explosion.

Major milestones took place in the history of sizing data volumes plus the evolution of the term Big Data. The following are some of them:

- In 1971, Arthur Miller stated in "The Assault on Privacy" that:

Too many information handlers seem to measure a man by the number of bits of storage capacity his dossier will occupy.

- In April 1980, I.A.Tjomsland gave a talk titled "Where Do We Go From Here?" at "Fourth IEEE Symposium on Mass Storage Systems" in which he says:

Data expands to fill the space available, I believe that large amounts of data are being retained because users have no way of identifying obsolete data, the penalties for storing obsolete data are less apparent than are the penalties for discarding potentially useful data.

- In 1997, Michael Lesk publishes "How much information is there in this world?" in which he concludes that:

There may be a few thousand petabytes of information all told, and the production of tape and disk will reach that level by the year 2000. So in only a few years, (a) we will be able to save everything- no information will have to be thrown out, and (b) the typical piece of information will never be looked at by a human being. (https://www.forbes.com/sites/gilpress/2013/05/09/a-very-short-history-of-big-data/2/#1c3097c24343).

The term Big Data was coined in 1998 by Mr. John Mashey, Chief Scientist at SGI. Even though Michael Cox and David Ellsworth seem to have used the term 'Big Data' in print, Mr. Mashey supposedly used the term in his various speeches and that's why he is crediting from coming up with Big Data. But some various sources say that the first use of the term Big Data was done in an academic paper- Visually Exploring Gigabyte Datasets in Realtime(ACM) (OECD, 2015; Mark A. Beyer & Douglas Laney, 2012).

The following are the differentiators of Big Data over Traditional Business Intelligence solutions:

- Data is retained in a distributed file system instead of on a central server.
- The processing functions are taken to the data rather than data being taken to the functions.
- Data is of different formats, both structured as well as unstructured.
- Data is both real-time as well as offline data.
- Technology relies on massively parallel processing(MPP) concepts.

The Big Data challenges include capturing data, data storage, data analysis, search, sharing, transfer, visualization, querying, updating and information policy. Organizations have to compromise and balance against the confidentiality requirements of the data. Organizations must determine how long the data has to be retained. With the advent of new tools and technologies to build big data solutions, availability of skills is a big challenge for CIO's. A higher level of proficiency in the data science is required to implement big data solutions today because the tools are not user-friendly yet. (Bill Franks, 2012).

Analogous to the Cloud Computing architecture, the Big Data landscape can be divided into four layers.

- **Infrastructure as a Service (IaaS):** This includes the storage servers, and network as the base, inexpensive commodities of the big data stack.
- **Platform as a Service (PaaS):** The unstructured data stores and distributed caches that can be logically queried using query languages serves the platform layer of Big Data.
- **Data as a Service (DaaS):** The tools required to integrate the PaaS layer using search engines, integration adapters, batch programs and so on is housed in this layer.
- **Big Data Business Functions as a Service (BFaaS):** Industries like health, retail, e-commerce, energy and banking build applications that serve a specific purpose and hold the DaaS layer for cross-cutting data functions.

Recent Gartner's "Hype Cycle for Emerging Technologies", visualizes the absence of Big Data. Big Data got a permanent place in the Gartner Cycle for the past 5 years in the emerging technologies, but shockingly it was removed in July

2016. The reason was given by Burton, Gartner Analyst that *I would not consider big data to be an emerging technology. This hype curve is very much focused. I look at emerging trends.* (http://www.gartner.com/technology/research/hype-cycles/).

Here are some of the big data providers that are offering solutions in the specific industries:

- The Securities Exchange Commission (SEC) is using big data to monitor financial market activity. They are currently using network analytics and NLP to catch illegal trade activity in financial markets.
- In Communications, Media and Entertainment, industry simultaneously analyze customer data along with behavioral data to create detailed customer profiles.
- Some hospitals are using Big Data techniques to use evidence-based medicine as opposed to administering several medical/lab tests to all patients who go to the hospital.
- Big Data has also been used in solving today's manufacturing challenges and to gain competitive advantage among other benefits.
- In public services, Big Data has a very wide range of applications including energy exploitation, financial market analysis, fraud detection, health related search and environmental protection.
- In the field of Insurance, Big Data is used to provide customer insights for transparent and simpler products, by analyzing and predicting customer behavior through data derived from social media, GPS- enabled devices and CCTV footage.
- Smart meter readers allow data to be collected almost every 15 minutes as opposed to once a day with the old meter readers. This granular data is being used to analyze consumption of utilities better which allows for improved customer feedback and better control of utilities use (David R. Hardoon & Galit Shmueli, 2013).

We have 4 V's of Big Data:

- **Velocity:** The data is increasing at a very fast rate. It is estimated that the volume of data doubles every year.
- **Variety:** Now a days data is not stored in rows and columns i.e., structured format. We see data is being stored in the form of log files i.e. unstructured.
- **Volume:** The amount of data which we deal with is of very large size of peta bytes.
- **Veracity:** Explains the reliability of data (Foster Provost & Tom Fawcett, 2013).

2. BIGDATA MODELING (4 V'S)

Bigdata can be defined as a collection of large amount of data sets. Using traditional tools it is very difficult to handle such a vast amount of datasets. This is the reason that bigdata become important in storing and analyzing large amount of data. These data can be mined by using data mining tools or DBMS tools. Based on this the main components of big data can be termed as set of 4 V's (P. Hitzler & K. Janowicz, 2013).

2.1 Volume, Velocity, Variety, Veracity

- Volume refers to the amount of data
- Velocity refers to the rate at which data can be generated and transmitted
- Variety refers to the different types of data such as structured, semi structured and unstructured
- Veracity refers to the integrity of the data

The bigdata differentiates from the traditional data based on these 4 components:

Figure 1. Modeling of Bigdata

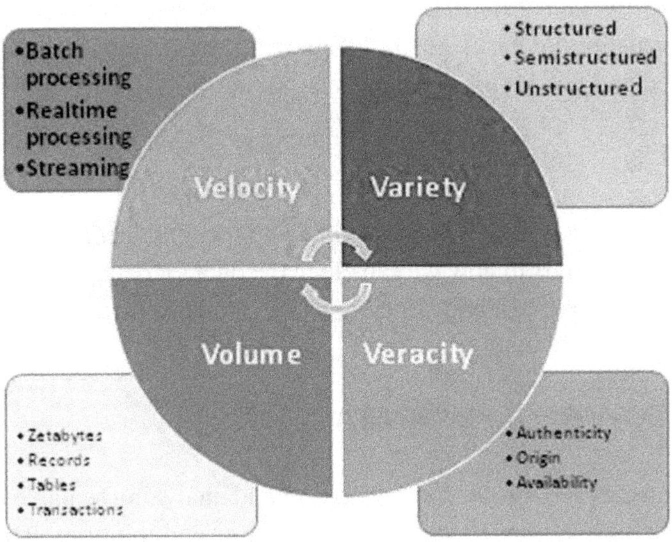

2.1.1 Volume

We live in the data age and in 2013 it was analyzed that the total volume of data storage is 4.4 zeta bytes and in 2020 it will become 44 zeta bytes (1 zeta byte is 10^{21} Bytes. These very much amount of data can need to be stored and analyzed. These data are from different sources and also need to be combined for better results.

2.1.2. Velocity

The data is from multiple sources and these sources are to be run parallel. Hence the important next issue is how to run these sources parallel with very high speed for data generation and transmission. For example consider a weather sensor which collects the weather data from multiple sources in each and every hour. These data is need to be move on to a particular storage and this data log is very high. Traditional systems are not capable of doing this storage and frequent movements.

2.1.3. Variety

The data sources can be of different types which can be collected from weather sensors, social networks, stock exchange and smart phones. The data includes text, images, audio, video or any other data logs. The data can be classified mainly into three such as structured data, semi structured data and unstructured data. Traditional distributed systems such as RDBMS, volunteer computing and grid computing can handle only structured data. Bigdata differs from these systems by it can handle semi structured and unstructured data also.

2.1.4. Veracity

The bigdata can handle huge amount of data and this data need to be correct also. Hence the Veracity refers to how we can clean the data for data preprocessing stage. The data need to be relevant and valuable (X. L. Dong & D. Srivastava, 2013).

3. ARCHITECTURE OF BIGDATA

Bigdata is treated as set of tools for developing and analyzing scalable, reliable and portable data. It serves as key for design infrastructure and solutions. It interconnects the existing and organizing the resources and consist different layers such as:

- **Bigdata Sources:** The location which produces the data.
- **Messaging and Storage:** The facilities where the data is stored.
- **Bigdata Analysis:** The different tools for analyzing the different types of data.

A bigdata architecture is designed so that it can handle processing, storage and analysis of complex large data .Bigdata can handle processing of big data sources, real time processing of data, predictive analytics using machine learning approaches and exploration of interactive data (K.Bakshi, 2012).

The architecture includes the following components:

- **Data Sources:** Real time data sources, application of data sources such as relational databases. Static files produced by webservers and other applications.
- **Data Storage:** Bigdata is not the first distributed processing systems. But it can store high volumes of data known as Data Lake, than other traditional systems. Bigdata can prepare the data for analysis and then these analytical data store can be used to serve different types of queries. The analytical data store can also provide metadata abstraction and low latency NoSQL technologies.
- **Processing of Data:** Bigdata provides interactive and batch processing including real time applications. Bigdata solutions process the data files using batch systems to filter and aggregate. If the solutions include real time sources, the bigdata architect can include stream processing also. The processed stream data is then written into an output sink (B. Ramesh, 2015).
- **Service Orchestration:** Most of the bigdata solutions consists of repeated data processing operations, then transform the encapsulated source data. The movement of data between multiple sources and destination and load the processing data in to an analytical store is an important issue in traditional systems. This can be automated by using service orchestration in bigdata by using the tools such as sqoop and Oozie.

4. TOOLS USED IN BIGDATA ANALYTICS

4.1 Big Data Analytics with GIS Datasets

A number of open source tools, frameworks and query languages have been introduced to analyse big data. MongoDB is a famous NoSQL based data analytics

Figure 2. Architecture of Bigdata

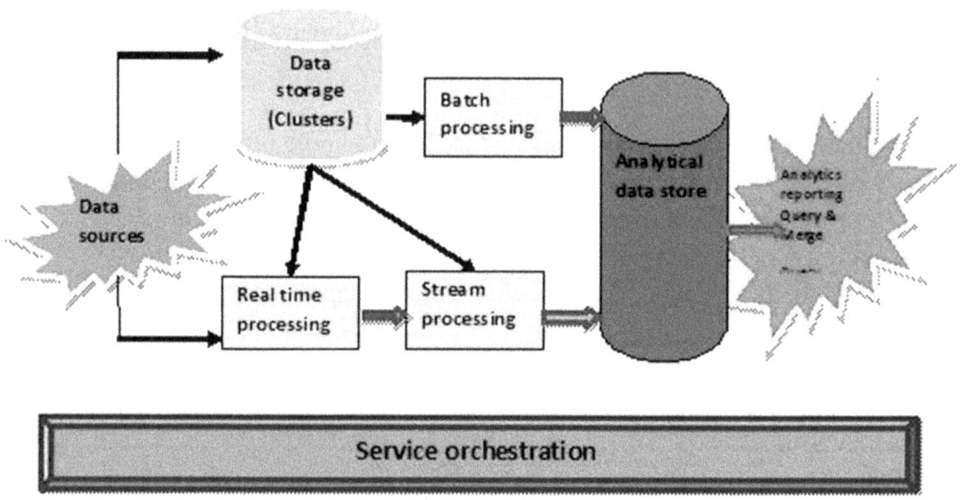

tool that provides option to visualize, analyse and explore datasets. In this section let us explore a GIS dataset in Mongo Compass.

Here is the step by step implementation of the work:

Step 1: Install MongoDB Compass from mongodb.com as per the instructions provided in the website.

Step 2: Configure the MongoDB compass with the following details to get connected to the host.

Step 3: Add the 100YWeather dataset to the MongoDB Compass. The dataset appears on the right side of the screen. Expand the title to see the data collection. Click on the data collection to view the records inside.

Step 4: Document View. In mongoDB records are named as json documents. The screen lists a huge collection of weather data in 250,000 json documents, with 5 Indexes and size of 403MB.

Schema View

Schema view lists the attributes information and visualizes the data with data types.

To get the view of the Schema click on the Schema Tab at and top and then click on the green 'Analyse' button.

Figure 3. Configuration screen of the mongoDB Compass

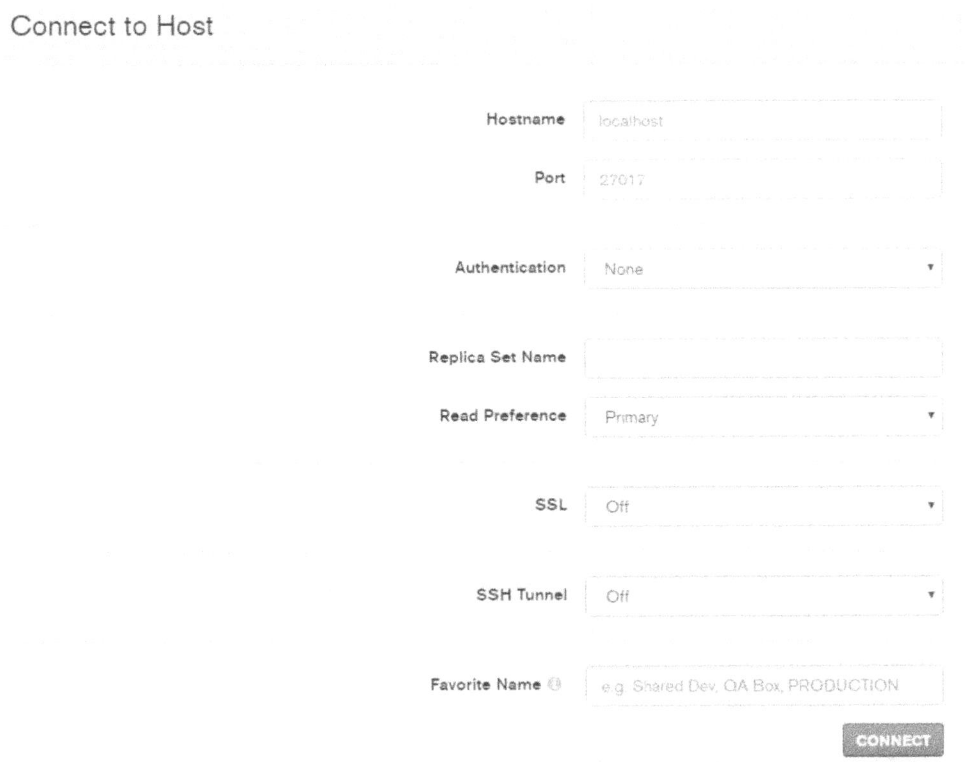

Figure 4. MongoDB Compass

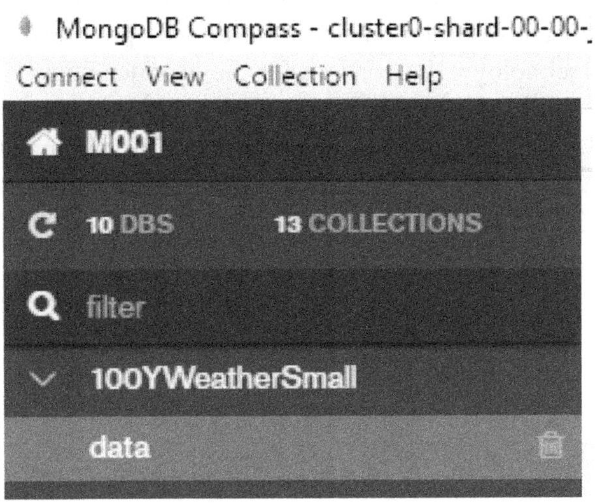

Figure 5. Document View

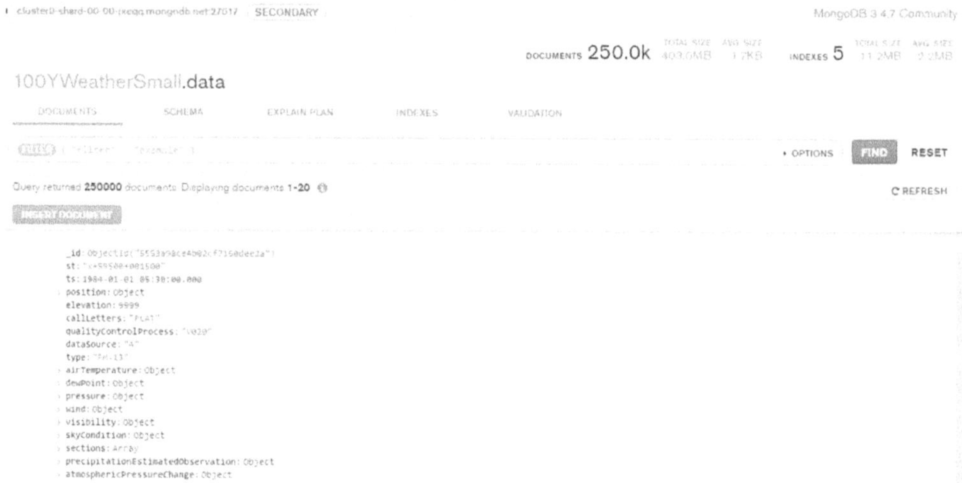

As shown in Figure 6, the query made to analyse the 100Y weather dataset has returned 250,000 json documents

The different tools in bigdata are summarized in Table 1.

5. CHALLENGES IN BIGDATA

The term 'Bigdata' implies that it is a big volume of Data. As many industries are generating data volumes from terabytes to petabytes, there developed a need to process the raw data in to useful information. In order to provide benefits to business and Information technology, the concern for the enhancement of big data storage and processing architectures has increased. Potentially Bigdata undergoes a lot of challenges with the characteristics, data analytics and processing capabilities of the system. In this section, we mainly discuss about the challenges in major research areas (X. Yi, F. Liu, J. Liu, & H. Jin, 2014).

Figure 6. Schema View

Table 1. Tools in Bigdata

Name of the tool	Type	Developed by	Functionalities	Task
Microsoft Azure	Paid	Microsoft	Big data analytics, cloud computing, machine learning, HD Insight	Platform
Amazon Web Services	Pay for service	Amazon	Big data analytics, cloud computing, machine learning, hosting	Platform
Hadoop	Open source	Apache	Distributed data processing, HDFS, YARN and Map Reduce	Framework
HBase	Open Source	Apache	Large scalable online data storage	Storage
Ambari	Web based	Apache	Hadoop Cluster Manager	Dashboard
Avro	Open source	Apache	Rich data types, RPC, Binary data format, dynamic code integration	Data serialization
Cassandra	Open source	Apache	Robust storage with Scalability, reliability, fault tolerance, decentralization	Storage
Pentaho	Pay for Service	Hitachi	ETL tools, No Sql, integration with Hadoop, mongodb and other services	Platform
Cloudera	Enterprise service	Cloudera	Enterprise level Hadoop services	Platform
Mongodb	Open source	Mongo dB	Json storage, realtime customization with sandbox, atlas and compass, data management, visualization, scalable, NoSQL query processing, Visualization, data wrangling, security	Real-time Data management
Talend	Open Source	Talend	Data integration, Data management and sandbox, scalability, agility, business integration	Data integration and management
Karmasphere Analyst and Studio	Pay for service	Karmasphere	Enterprise solution on clustered and unstructured Hadoop clusters, SQL, Map Reduce, Algorithm customization	Analytics
Skytree	Pay for sevice	Skytree	Highly accurate big data machine learning models, scalability, robust, automation, visualization	Advanced analytics and machine learning
Openrefine	Open source	Openrefine	Data pre-processing, scalable, works with messy data, Exploratory data analysis, ETL tools	Pre-Processing
Splunk	Pay for service	Splunk	User Behaviour analysis, Security, Machine data architecture, data integration and transformation	Analytics
Datacleaner 5.2	Pay for service	Neopost	Data pre-processing, anomaly detection, data cleaning, data health monitoring system, visualization, Standardization and profiling	Pre-Processing

continued on following page

Table 1. Continued

Name of the tool	Type	Developed by	Functionalities	Task
RapidMiner	Pay for service	Rapidminer	GUI, Machine learning, data pre-processing, predictive analysis	Machine learning
Weka	Opensource	University of Waikato	GUI, Machine learning, data pre-processing, Jython, feature selection, classification, clustering, association rule mining, data visualization, integration with spark, python, big data analytics	Machine learning and visualization
IBM SPSS Modeler	Pay for service	IBM	Text analytics, statistical analysis, decision making, data management	Data management and analytics
Oracle cloud	Pay for service	Oracle	Big data storage, Big data preparation, Compute, development, analytics, golden gates, processing, MySQL, Java support, Internet of things integration.	Platform
Apache pig	Open Source	Apache	Parallel data analysis, map reduce, scalability, optimization	Platform
Teradata	Pay for service	Teradata	Unified big data architecture for enterprises, business insights, analytics	Analytics
Datawrapper	Pay for service	Datawrapper	Line plots, bar diagrams, donuts, stacked bar, geographical data mapping, tables	Visualization
Blockspring	Pay for service	Blockspring	Data warehousing, data synchronization with apps	Data integration
Magento	Pay for service	RJ Metrics	Business intelligence, data visualization and exploration	Analytics
Ideata analytics	Pay for service	Ideata	Business intelligence, data preparation, knowledge discovery, self-service analytics, big data analytics and visualization	Analytics
Thingspeak cloud	opensource	Thingspeak	Data storage, sensor integration, real time analytics, visualization of graphs, integration of api with social media, apps	Analytics and visualization

5.1 Data Complexity

The Bigdata collects a large scale data from different sources and solve the computational problems. The most significant characteristics of big data are types of data patterns, different structures and complicated communication between data samples. Big data refers to the data (or) Information which can always be processed with advance technology like analytics, visualization methods and can also find hidden pattern in order to make an accurate decision rather than using additional compute system ("Global Data Center Traffic"). Many Business organizations have large amount of data but that whole data is used properly due to the lack of efficient systems so that the percentage of data utilization is keep on decreasing (*Oracle Big Data Strategy Guide*). Now the technology is growing day by day, the mobile devices and sensors playing an important role for generating data and then stored. For example, people can monitor working people away from office because the office people staying far away from office (*Cloudera's 100% Open Source Distribution of Hadoop*). For instance, railway companies decided to install sensors for every few feet in order to monitor internal and external event which causes train to meet accidents. By monitoring every event, railway people comes to know that which part is required to replace and which one is repaired (James Manyika, Michael Chui, Brad Brown, Jacques Bughin, Richard Dobbs, Charles Roxburgh, & Angela Hung Byers, 2011). The roads are equipped with sensors to read the information for reducing the chances of occurring natural disaster by analyzing and predicting generated data ("Big Data, Big Impact: New Possibilities for International Development"). Everyday few terabytes of data is generated from every business organization, to deal with the traditional methods such as information retrieval, discovery, analysis, sentimental analysis, but they are not fit for huge data. Currently, we don't have good knowledge on data distribution law and association relation of big data. Also lack of deep understanding association between data complexity and computational complexity arises. The lack of the processing methods in big data and all the other aspects confine our ability to implement new method and models to solve all these problems in big data. The basic problem is to understand essential characteristics of complexity of big data. Basic study on complexity theory will give clear view on complexity and how they are formed, how they are associated with other pattern. By getting clear understanding on complex theory, we can design and implement novel models to resolve problems of big data complexity.

5.2 Computational Complexity

Big data includes three main features such as fast-changing, multiple sources, huge volume . These features become difficult for traditional processing methods such as machine learning, data retrieval, data analysis. New big data computing tools are needed to overcome problems coming from independent and identical distribution of data for generating accurate statistics. For addressing the problems of big data, we require examining computational complexity and used algorithms. The storage available for generating data through social websites is not enough, because of the popularity of big data on storage and network. Server's offloading the resource intensive data into cloud and sending more data into cloud for processing. Offloading data into cloud does not solve the problem. But big data require getting all the data collected and retrieving from terabytes to petabytes. Offloading the entire data into cloud will take large time and also data is changing in every second which will make the data hard to be updated in real time. Offloading data from the storage location to processing location can be avoided by two ways: one is to process the storage location and second one is offload only required data which will take more time to process. Building indexes for storing data which will make easy the retrieving process and moreover reducing processing time. In order to address computing complexity in big data we need to understand about life cycle application of big data to study of centralized processing mechanisms which depends upon the behavior of big data. We need to get way from tradition approach of computing-centric to distributed computing paradigms. We need to focus on new methods and data analysis mechanisms for distributed streaming mechanisms. We are also required to focus more on boot strapping and local computation, new algorithms for handling large amount of data.

5.3 System Complexity

Big data can process high-volume of heterogonous data types and applications to support research on big data. As big data processing are not enough to handle generating high volume of data and real time requirements, these constraints pose to establish new system architecture, processing system and energy optimization model. Addressing the complexity problems will lead to designing novel hardware and software frameworks for optimizing energy-consumption on big data. Systems can process data that are all similar in size and structure, but make difficult to process when data is presented in different patterns and sizes. We need to conduct small scale research on all big data tools, different work load conditions, different data types, various data pattern, and performance evaluation in distributed environment, centralized environment, machine learning algorithm for performance prediction, energy optimized algorithms, energy consumption per unit and recursive work. We should focus on novel data processing systems which is able to process all kinds of data in different situations.

6. APPLICATIONS OF BIGDATA

- **Big Data Analytics(BDA):** This kind of application analyzes massive data using parallel processing framework. BDA applications use sample data in pseudo-cloud environment. After that they build in real cloud environment with more processing power and large input data. These applications utilize large data, which is unable to fit in hard drive. The data is generated from different sources like traffic, social websites, the online game information, and stock market and during international games.

- **Clustering:** User can easily identify group of people by using algorithms such as *k*-means algorithms through points and click dialog and based on specific data dimension. Clustering plays an important role in big data in order to address group of people by considering customer type patient documents, purchasing pattern, behavior products.
- **Data Mining:** The decision tree will help user to understand outcome and relation between attributes in expected outcome. This decision tree reflects the structure of that probability hidden in your data. Decision tree helps us to predict fraud risk, online registrations, online shopping, and disease risk.
- **Banking:** In banking sector, the use of sensitive data leads to privacy issues. Research shows that more than 62% of bank employees are cautious about their bank customer's information to privacy issues. Distribution of customer's data to different branches also leads to security issues. The investigation happened on banks data containing user's sensitive information such as earnings, savings, and insurance policies ended up in the wrong hands. This discourages the customers in sharing personal details in bank transactions.
- **Stock:** Data analytics can be used to detect fraud by establishing a comprehensive system in data base during private stock exchange.
- **Credit Cards:** Credit card companies depend on in-data base analytics to identify fraud transactions with accuracy and speed. This fraud transaction deletion will follow up users sensitive data such as amount, location and follow up before authenticate suspicious activity.

Enterprise

It will help the industry people exist around the world. Data doesn't have to move to work and back. It provides insight to business people to make accurate decision for less expensive than traditional tools.

- **Customer Goods:** A manufacturer of customer products gather data related to customer preferences and purchasing data of surveys, web tags, customer call centers, the text taken up from online web sites, everything that being said about product extract. By following this kind of analysis business people will come to know that which product is to be succeeded and others are to be failed. Finally, they can spot feature trends the products in the marketing media.

- **Agriculture:** Using sensors, agriculture firm collect the data of efficiency of crops. Initially, experts of agriculture field plant crops and run simulation to find how crops react in different conditions. Research on agriculture will collect data including various attributes, temperature, growth level, soil composition, in order to find optimized environment for specific gene types.
- **Finance:** The popular financial companies are focusing more on credit scoring using third-party. Today financial institutions using their own analysis for generating credit score for existing users using wide range of information such as credit card, investments, transactions, savings.
- **Economy:** Hadoop can assist organization to perform low cost transactions.
- **Telecom:** People carry face reorganization technology in their pocket. Android users use remember app to recognize (or) collect data related to that snapped image from data base, while "I phone" users unblock their devices with recognize me, this app developed widely to save nearly $2.5 million per year for solving forgotten –passwords.
- **Health Care:** Traditional health care industries has lagged behind than the other industries in big data, every stack holder of hospital taking decision independently rather than depending upon big data tools (James Manyika, Michael Chui, Brad Brown, Jacques Bughin, Richard Dobbs, Charles Roxburgh, & Angela Hung Byers, 2011).

Most health care stakeholders invested in information technology because of accurate results. The traditional systems have limited ability to cause data become more certain. The healthcare industries itself create issues. It becomes difficult to share data among department even within hospital. The important information of payee, pharmaceutical company will be available within single department because organizations lack of knowledge on integrating data and result obtained. Big data advances healthcare ability to work with enormous data even though data is presented in different formats.

7. CONCLUSION

Bigdata is an important technology in our data era which can handle structured, semi structured and unstructured data. It provides a viable solution for large and complex data and become a challenge in nowadays. Each bigdata system provides a massive power and for providing this different tools are used. This paper presented the important aspects of big data, architecture and its applications. However the challenges are also presented in this paper. It is clear that now we are starting in the bigdata era and we have to discover many things about big data for competing this data world.

REFERENCES

Bakshi, K. (2012). Considerations for big data: Architecture and approach. *Aerospace Conference*, 1–7. 10.1109/AERO.2012.6187357

Big Data, Big Impact: New Possibilities for International Development. (n.d.). World Economic Forum.

Bill Franks. (2012). *Taming the big data tidal wave*. Wiley.

Cloudera's 100% Open Source Distribution of Hadoop. (n.d.). Retrieved from http://www.cloudera.com/content/clou dera/en/products/cdh.html

David, R. (2013). *Getting started with business analytics – insightful decision making*. Talor & Francis Group.

Dong, X. L., & Srivastava, D. (2013). Big data integration. In *Data Engineering (ICDE), 2013 IEEE 29th International Conference on*, 1245–1248. 10.1109/ICDE.2013.6544914

Global data center traffic – Cisco Forecast Overview. (n.d.). Retrieved from http://www.cisco.com/en/US/solutions/ collateral/ns341/ns525/ns537/ns705/ns 1175/Cloud_Index_White_Paper.html

Hitzler, P., & Janowicz, K. (2013). Linked Data, Big Data, and the 4th Paradigm. *Semantic Web*, *4*(3), 233–235.

Manyika, Chui, Brown, Bughin, Dobbs, Roxburgh, & Byers. (2011). *Big data: The next frontier for innovation, competition, and productivity*. McKinsey Global Institute.

Mark, A. (2012). *Beyer and Douglas Laney. "The Importance of 'Big Data': A Definition*. Gartner.

OECD. (2015). *Data-driven innovation: big data for growth and well-being*. Paris, France: OECD Publishing.

Oracle Big Data strategy guide. (n.d.). Retrieved from http://www.oracle.com/us/technologies /big-data/big-data-strategy-guide- 1536569.pdf

Provost & Fawcett. (2013). *Data science for business*. O'Reilly.

Ramesh, B. (2015). Big data architecture. In *Big Data* (pp. 29–59). Springer. doi:10.1007/978-81-322-2494-5_2

Yi, X., Liu, F., Liu, J., & Jin, H. (2014). Building a network highway for big data: Architecture and challenges. *IEEE Network*, *28*(4), 5–13. doi:10.1109/MNET.2014.6863125

Chapter 9
Big Data Processing:
Application of Parallel Processing Technique to Big Data by Using MapReduce

Abhishek Mukherjee
VIT University, India

Chetan Kumar
VIT University, India

Leonid Datta
VIT University, India

ABSTRACT

This chapter is a description of MapReduce, which serves as a programming algorithm for distributed computing in a parallel manner on huge chunks of data that can easily execute on commodity servers thus reducing the costs for server maintenance and removal of requirement of having dedicated servers towards for running these processes. This chapter is all about the various approaches towards MapReduce programming model and how to use it in an efficient manner for scalable text-based analysis in various domains like machine learning, data analytics, and data science. Hence, it deals with various approaches of using MapReduce in these fields and how to apply various techniques of MapReduce in these fields effectively and fitting the MapReduce programming model into any text mining application.

DOI: 10.4018/978-1-5225-3643-7.ch009

INTRODUCTION

In the steadily changing perception of information technology, data being gathered and looked through for business knowledge purposes have achieved excessive levels. Welcome to the big data revolution. Big data is the term used for data set so large or complex such that they cannot be modified or processed in the conventional programming environment or softwares for generating specific prediction or output thr, i.e., the day to day data processing application softwares are not able to deal with them. This happens because the day to day software systems have limitation of their process and also these huge chunks of data contain so many outlier and exaggerated information that they are cleaned and made fit for the processing before they are dealt with. For processing, curing, storing, analyzing, decision making, transferring, visualizing, query processing, updating using this huge hunk of data, the Big Data environment is used.

Now we live in such a generation when huge chunks of data are generated at every moment which is used by the business persons for analysis. If these data are to be used for processing, then serialized way of processing will not give result efficiently because that will consume time more than required. So, to avoid this problem of the serialized processing, we go for parallel processing in Big Data environment. This parallel processing helps us to reduce processing time because the data sets are dealt in relatively small parts and they are merged after that.

This is the brief idea about the chapter and we will see more details inside the chapter.

BACKGROUND

While defining Big Data, it is very difficult to differentiate exactly between data and Big data. The data can also be processed in this environment but that is not suitable way for that. But big data handling is not at all suitable in the day to day software system. In processing the Big data, the input data are having the qualities like- Volume: The quantity of the data determines the weightage and potential insight. Variety: The nature and variety of the data analyses how eclectic the data is. Velocity: The data actually varies with time with variance and this determines how useful the data is.Variability:Inconsistant data set of the base analysis can hinder the handling.Veracity: The quality of captured data can vary greatly, affecting accurate analysis. These qualities can be considered as formal definition of the Big Data.

While dealing with the parallel processing, Parallel computing (Kumar, V., Grama, A., Gupta, A., & Karypis, G., 1994) is the specific type of computation in any computing or processing environment in which many calculations or the execution of processes are done simultaneously, i.e., large problems can be divided into smaller problems for processing and then at the end, they are merged for producing output. Parallel processing reduces the processing time drastically because more than one portion of the data are being processed at the same time.

MAIN FOCUS OF THE CHAPTER

The main focus of this chapter is to know about the modern day systems to keep up and process all this information in a timely manner, new technologies have been developed and old ones have been improved upon. Massively parallel processing, or MPP (Metropolis, N., 1986) and MapReduce are such technologies.

MPP manages the coordinated processing of huge datasets utilizing different processors. This empowers fast rates of execution for many-sided inquiries running against extensive information distribution centers. The primary reason this technology came about is a result of the huge amounts of data that were inundating applications not intended for such huge volumes. The need to process this data for the reasons for investigation urged originators to create ultra-quick preparing systems. Without the methods MPP employs, a query may take very long time to finish, making present day business insight frameworks and data warehouses not as much as helpful. MPP is at the heart of a wide range of sorts of huge data solutions, and has made considerable advances as a critical innovation. Amazon Redshift,is a fast, fully managed, petabyte-scale data warehouse solution and a prevalent cloud-based information (Gupta, A., Agarwal, D., Tan, D., Kulesza, J., Pathak, R., Stefani, S., & Srinivasan, V., 2015) warehousing arrangement, utilizes MPP design to accomplish to a great degree quick inquiry execution. MPP is one of the five key execution empowering agents of Redshift, alongside columnar data storage, data compression, query optimization, and compiled code.

What MPP does is genuinely straightforward in principle. It separates large, hard to manage datasets into effortlessly workable chunks, assigning each to a processor. There could be upwards of hundreds or even thousands of processors dealing with chunks from the same dataset. When all the data has been processed, results from the many different processors are combined for a final result set. How MPP really accomplishes this is through an informing interface the individual processors use to communicate with each other. Every processor has its own working framework and memory. This permits everyone to work autonomously, on its assigned section of the database, while loosely communicating with alternate processors. It is hence that MPP frameworks are known as a “shard nothing” framework. Since OS, memory, disk or whatever else is not being shared among the individual nodes in a shared nothing design, conceivable outcomes and points of interest for fine-tuning the framework can be realized. The MPP framework can be scaled to the same number of extra nodes as required, isolating the work further and accelerating the framework to optimum levels. Additionally, since the processor of every node works independently of the others, there is no bottleneck to hinder performance.

MPP is not by any means the only technology available to encourage the handling of substantial volumes

of data. MapReduce, a part of the Apache Hadoop Venture, is another innovation that accomplishes similar things MPP does, however with a few contrasts. Actually, you may even say MPP and MapReduce are distant cousins. At first glance, technological differences between the two may not appear to be too far separated, but rather relying upon the requirements of your data distribution center, picking one over the other could have a tremendous effect.

- **Performance:** The optimization and distribution components of MPP permit it to deal with the distribution of data among the different nodes. This rates up processing time considerably, making it a better performance decision over MapReduce. Additionally, MPP databases fit in with the ACID compliance which remains for atomicity (Coyle Jr, D. J., Chang, A., Malkemus, T. R., & Wilson, W. G., 1997) i.e. rollback of all updates is required if the update of any node fails, consistency (Eswaran, K. P., Gray, J. N., Lorie, R. A., & Traiger, I. L., 1976) i.e. the data satisfies certain consistency constraints, isolation[14] i.e. isolation requires that transactions observe the" latest" snapshot of the database, and durability (Elnikety, S., Dropsho, S., & Pedone, F., 2006) i.e. commits and making the effects of transactions durable are performed in one single action. ACID ensures database transactions are processed accurately and reliably. This is not something that is naturally implemented in Hadoop, giving MPP the general execution edge over MapReduce.
- **Scaling:** The highly specialized hardware used by MPP frameworks make adaptability a difficult and costly recommendation. MapReduce and Hadoop, be that as it may, can be deployed to economical commodity servers, permitting the groups of nodes to develop as required.
- **Deployment and Maintenance:** MPP is generally simple to deploy and maintain. Hadoop and MapReduce, on the other hand, can end up being a major implementation project requiring expensive and specialized expertise that may not be available in-house.
- **Data Restrictions:** Unlike MPP, Hadoop and MapReduce can tackle unstructured data without the need to process it before use. There is no need to massage the data before it can be used. So, MapReduce is preferred slightly more here.

- **Language:** The language behind MapReduce's control component is basically Java. MPP uses SQL, making it easier to use and more cost effective. SQL is an outstanding query language, for the most part used by database experts, wiping out the need to procure expensive Hadoop experts. Additionally, existing SQL-based business intelligence tools are supported with MPP. This is not the situation with MapReduce, and alternative solutions must be explored.

Is MPP better than MapReduce or vice versa? This depends on the goals of each organization, for them they are different tools that suit different situations. As a matter of fact, there are some organizations that use both MPP and MapReduce, affording them the advantage of the best of both worlds.

ISSUES, CONTROVERSIES AND PROBLEM

- **There is No Need to Distinguish Big Data Analytics From Traditional Data Analytics.**
- **Hadoop is Not Always the Best Tool:** There are situations when Hadoop is not adequate for big data in organizations; while crunching real time data, when the organizations have numerous data sources and business processes that can't be fit into a single data infrastructure and when the organizations do not have the ability pool for complicated data science tools like MapReduce.
- **Size of the Data Does Not Matter as in Real Time Analytics, Data May be Changing. What is More Important is its Recency:** Big data may not be the right choice, while crunching real time data, because Hadoop like tools process data in nodes. With a temporal measurements varying over time, the quantity of data doesn't matter unless it ties in with time. What really matters is the sample size at a particular instance (which is again a dimension) and not the overall sample size.
- **Bigger Data are Not Always Better:** Data adequacy plays a vital role when we run samples across various dimensions. The quality of data used for crunching makes the quality of insights. If the signal to noise proportion is high, the exactness of results may shift for dirty samples
- **Big Data Involves an Ethical Issue:** Security. In the era of big data, the debate between privacy and personalization will be progressing. Big Data is a major ordeal today however privacy guidelines is similarly essential. It is unethical if people are unaware that their data is analyzed.

Figure 1. Number of line vs processing time graph

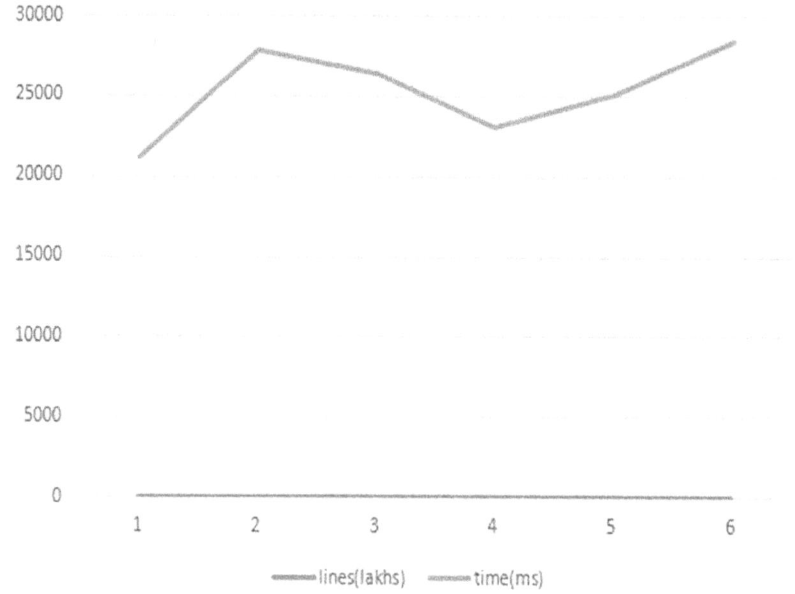

Figure 2. 5 lac line processing

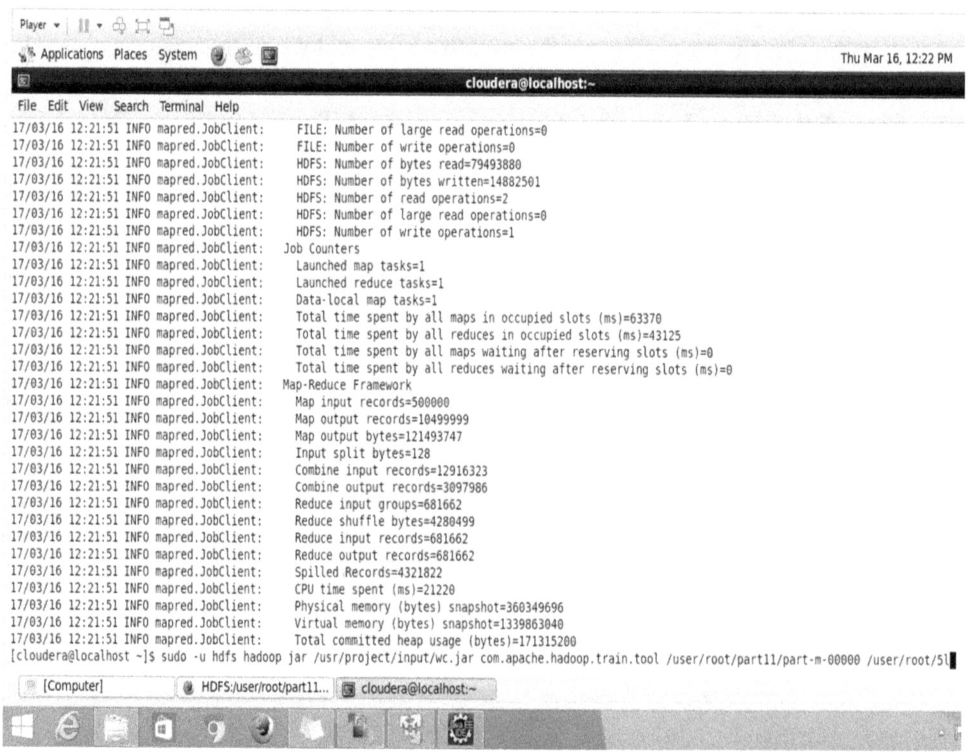

Figure 3. 10 lac line processing

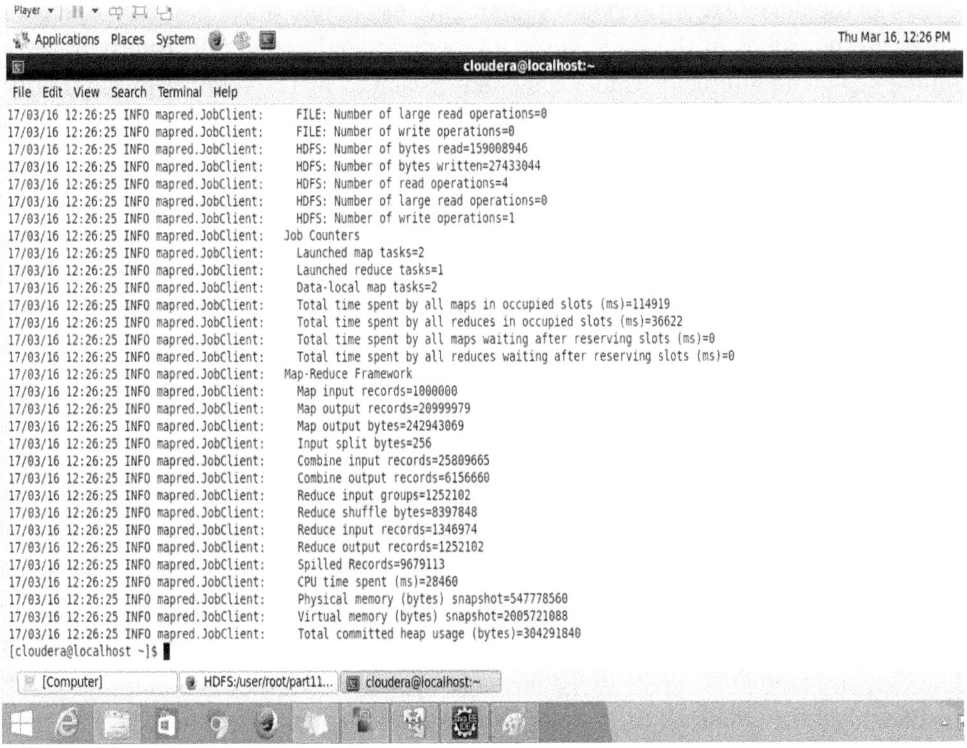

SOLUTIONS AND RECOMMENDATIONS

All the problems stated above have various solutions one of them being map reduce implementation of various analysis algorithms. But before this we need to know about the phases of map reduce algorithm and how it works, the data flow among the various phases and how it deals with processing the data parallel. Map reduce is a software framework developed and used by Google™ (Schatz, M. C., 2009) to support parallel processing tasks explicitly for intensive data handling. It serves as a data warehouse to handle jobs parallel and permit large scale distributed data analysis (Taylor, R. C., 2010). Each task in Map Reduce is broken into the following main phases which are mapper, sort and shuffle and finally reducer. A phase named combiner which can be used as a localized reducer is also used between mapper and reducer for optimization purposes i.e. to send lesser amount of data across the network. The output of the map tasks, called the intermediate keys and values, are sent to the reducers. The processing of data takes a collection of input key/value pairs, and produces a set of output key/value pairs. As stated above the user decides the process flow according to the map and reduce tasks where the map task basically aims to initially create the key and value for later processing as required by the further stages. In the intermediary sort and shuffle phases operate like a group by operator where all the values of the same key are grouped together and thus finally it gives unique keys with a tuple of values attached to each of it. The reducer phase finally does the work of giving a key and value pair separated by a delimiter where the value logic can be thought of as giving a single value from a tuple or list of values as per the requirement of the program. As stated above many analysis tasks can be done by the usage of map reduce some of which will be aggregation, data filtering, data organization and finally a few machine learning algorithm (non-iterative). This mainly deals with getting insights about the various statistical parameters of data which are generally used to get various information about the general nature of data variation, data summarization and range. A few algorithms of each category are listed below

Mean: The arithmetic mean, is the measure of central location of data. In the cases of a robust measure of central tendency will often provide a better estimate of the mean of the data (Tukey, J. W., 1977). It can be measured by summing up all the values of given set and dividing the sum by the number of values. Hence we need to fit our calculations according to the map reduce algorithm so that we can form tuples meant for mean calculation from a set of data. Here we deal with numeric data only as summation of string values is not possible. Mean calculation needs to assume a key for which all the values will be accumulated and divided by number of terms. Here we will assume a dataset which contains the values separated by ',' containing the numbers separated by ',' which aims at finally finding the average value of each column. Hence we will assume two phases initially(map, reduce):

1. **Map Phase:** As stated above the map phase reads the input file line by line and hence the key and value pairs for the input of map phase serves as the byte offset of each line as the key and the line as a whole as the value associated with each key i.e. the byte offset. Usage of byte offset address translation may be completed in as few as a single clock cycle (Senthil, G., 2004). Hence this serves as a benchmark for iterating through the line and performing required operations. Thus while writing the code we can assume a variable which is retained in every call of the iterator function and hence is can be programmed as a counter which is incremented as we keep getting the values for any particular column. Hence map output will have many keys associated with many values corresponding to number of lines in the input file. During the sort and shuffle phase all the values i.e. the numbers in the same column are sorted according to the key and finally the output of this phase is key associated with a tuple(list) of values which means that each column number(key) is associated with all values corresponding to all the values in their respective columns.

2. **Reduce Phase:** As stated above the input to the reduce phase is the column number as a key and a list of values as the values associated to the key. Hence now the reducer phase has access to all values in a particular column as a list. Hence a running sum is to be maintained which retains its value as long the iteration of the list of values takes place which keeps adding the numbers and hence at the end of the iteration we have the final sum with us in the variable. Thus the sum obtained in divided by the number of values of the list and is finally associated to the input key i.e. the column number. Hence the value of mean is obtained in this way for each of the column.

This method may sound perfect but it has a huge flaw. Consider a huge dataset with a large number of lines. Hence when the reducer is fed in with the list of values each key will have as many values in its tuple set as the number of lines in the data. This will increase the network traffic i.e. a large number of values will need to serialized altogether at the end of sort and shuffle and will need to be sent to the reduce phase. This can be optimized by the diving the file by the basis of size/number of lines and by usage of the combiner phase which will serve as a localized reducer here. To make it efficient storage of two things in each value i.e. the average of each key and the number of terms for which the average has been calculated. Thus instead of sending all the values associated to the columns one can find the average of a subset of the columns and send the average accompanied by the number of terms involved in calculating the sum. Hence at the reducer end a list of values (depending on the number of pieces the file was divided into) are obtained. Now to calculate the average each of the value obtained for each key is multiplied with number of occurrences which is present in the tuple. Summing up of all such values and division by the sum of number of occurrences produces the mean of the data and hence optimization is achieved.

KNN(k nearest neighbor): It is a very famous classification algorithm used to specify particular classes for specific data available. KNN shows the maximum accuracy as compared to the Naive Bayes and a few other classification based algorithms (Bijalwan, V., Kumar, V., Kumari, P., &Pascual, J., 2014). T. Hence since it classifies data it must use some metric for performing classification operation. In this case we take Euclidian distance as the metric based for the classification task. The basic algorithm follows the principle of taking the parameter 'k' as input and finding all the classes of data which are sorted by their distance from the concerned data point. After the classes of data are sorted by distance the first k classes are taken in ascending order. Finally the occurrence of each unique class is calculated. The concerned data point belongs to the class which has maximum occurrence in these obtained 'k' nearest classes. Since it is used as a non-iterative algorithm here we can frame phases according to the map reduce framework for implementing knn algorithm.

1. **Map Phase 1:** In the first phase of the implementation which is a mapper phase the data set is iterated to find the distance of each of the data present in the input data set from the data value whose class variable is to be calculated. Each of this distance has an associated class variable i.e. the class to which the particular input data value from which the distance has been calculated belongs to. The usage of a tree map comes handy here. Implemented as a linked "binary tree" structure, very fast: O(log N) ; keys are stored in sorted order (Reges, S., &Stepp, M., 2014).Tree map is basically a modified form of a binary search tree which stores certain data elements according to associated values which are generally numeric values. This is because the values stored in binary search tree are stored by comparing with the available values already stored previously. This data structure is used so as to access and store each element in a sorted fashion in an easier manner. Since the worst case complexity of finding and storing any element in O(n) this a tree map is preferred. In this case the value stored in the class variable and the key(based on which the class variable is stored) associated with it is the distance associated to it. Since the value of 'k' is static for the process as soon as the size of the tree map exceeds 'k' the value with the largest distance is eliminated. This helps in storing only the nearest 'k' class variables
2. **Map Phase 2:** The previous phase did the main task of storing the nearest k class variables. This phase is very similar to the word count map phase where each class variable is associated with '1' as a value. Hence at the end of this phase we have as key and value pairs with class variable being the key and 1 as the value associated with the key
3. **Reduce Phase 1:** In this phase the sort shuffle outputs give key as each unique class variable of the 'k' nearest data elements and '1' as the tuple of values associated with these key elements where '1' occurs as many times as the occurrence of the class variable obtained from the map phase. In this phase all the values associated with each key is added so finally the output of this phase in each unique class variable as the key and the number of times each class variable occurs in the 'k' nearest set.
4. **Map Phase 3:** This is a phase which gives the final output of the class to which the data tuple belongs. Here again the tree map is used to store the largest key i.e. the key occurring the most number of times.

Data Recommender System: A key building block for collaborative filtering recommender systems is finding most occurring data in an input dataset (Ristoski,

P., Mencía, E. L., & Paulheim, H., 2014). This algorithm aims in tagging documents which is basically based on the fact of how often any particular word occurs in a document. This is especially useful for finding keywords in a document and important parts of a document. Hence it gives useful insights into data to find the important contents of input data. The initial phase involves the word count algorithm. In this phase mainly the words in the file as segregated from each other to carry out the frequency aggregation task for the input file. In the next few phases the data is sorted on basis of the frequency of each word as obtained. The detailed analysis and phase wise tasks are explained as follows:

1. **Map Phase 1:** In this phase the data is read line by line and all the words are collected in the forms of tuples where each words is linked with '1' which in a way means the frequency of occurrence of that words in the file and hence this phase mainly identifies all the words in the input file and hence the main task of this phase is to identify all the words irrespective of their frequency of occurrence in the input file and hence this constitutes the first phase of the job.
2. **Reduce Phase 1:** This phase involves the output of the map shuffle and sort phases. The shuffle and sort phases do a group by operation on the key i.e. the words in the file which are obtained from the map phase. The sort phase collected all keys(words) of same value together and hence by the use of string comparison all the keys are sorted. In the shuffle phase all the values of the same keys are collected together to form a tuple of values (1's) which effectively results in the number of 1's associated with each key as the number of times the word actually occurs in the file. Finally in the reduce phase all the 1's of each key are added together and hence the frequency of each unique key is obtained by summing up of the values.

At the end of these two phases we have the frequency of each word occurring in the input file and hence we get the output of this phase as the words was keys and the frequency of occurrence of each word as the value. This thus makes the keys as unique. Now for the final output the words i.e the keys need to be sorted on the basis of their frequency obtained.

3. **Map Phase 2:** In the next steps the advantage of sort phase is taken hence the values i.e. the frequency will now be used as a key and the word which was the key earlier will now be used as a value. Hence the task of this mapper is to exchange or swap the frequency as the key and the word as the value
4. **Reduce Phase 2:** During the sort and shuffle phases the words with same frequency are collected together in the form of tuples and hence the output

of sort and shuffle phases will be sorted frequency of words as key and all the words with the same frequency together. Since reducer outputs single key and value pairs instead of key and list of values which may occur in the case of many words having the same frequency and. Hence when the iteration across the list of values for the key i.e. the frequency occurs an empty string is assumed initially which is concatenated as and when new words are obtained from the list of values as obtained by the reducer. This results in the formation of key as the frequency and the words which have same frequency as the value. Special characters can be used as delimiters between the unique words so that separation is obtained between the words of same frequency are obtained. These phases can be optimized by the usage of combiner phase between map phase 1 and reduce phase 1 so that less amount of data is to transferred across the network hence optimizing network traffic usage. The combiner serves as a localized reducer by counting the frequency of all words in pieces of data which can later be combined easily since it is commutative operation.

Ordered Sorting: The aim of this algorithm is to perform a optimized sorting of the input dataset in a parallelized manner. So basically it aims at sorting data based on a parallel key. Sorting is easy in sequential programming. Sequential programming based sorting is easy but since map reduce uses a 'divide and conquer' approach it becomes a bit difficult to make the process work in the map reduce programming model. One approach may be to take all the input data and sort data on particular parts of the data but that will not work as a global approach since sorting is not a commutative and associative property. The type of data flow is that partitions are required so that each partition is sorted locally and when all these partitions as concatenated we get the data sorted as a whole Sorted data has a number of useful properties. Sorted data provides many insights into data by giving the range and timeline of data. Sorted data also helps in performing search operations on data sets by giving binary search type methods which is O(logn) complexity as compared to normal searching methods which mostly have O(n) complexity. Hence the applications range from indexing data and finding some data within specific ranges in a very efficient and optimized manner. The main requirement of this is that the key should be comparable type of data so that the sorting can be performed. This is performed by the usage of keys which denote a specific section of data i.e. keys as so constructed that data is evenly distributed among them and only data within specific ranges occur within each partition. This will hence require a pre analysis of data so as that the map reduce job gets to know about the data variation as a whole and hence can construct keys for the partition operation. This will then later help in assigning data to each of these partitions each of which can be sorted locally and later the partitions can be sorted according to the constructed key so that at the end we have the sorted data. The analyze phase may be optional for some specific types of data like the datasets which have a unique element in the data which can hence be used to find the exact range of a particular parameter in data and hence the data can be partitioned on the basis of that parameter very easily. The analyze phase if required samples data randomly and hence the partitions are built over those samples. The mapper does a simple random sampling. When the output is given only the key which is the sorted value is given as output with the value associated with it as a null since the value will not be associated to any other value. Only one reducer is used here since that will come as a final sorted list and hence data handling will be easy in that manner. Finally the practitioner phase will be executed for data range as obtained. The mapper extracts the sort key in the same way as the analyze step. However, this time the record itself is stored as the value instead of being ignored. The custom practitioner takes in all values and sends to reducer based on the ranges as decided by the analysis phase. The reducer just outputs the values as obtained as the sorted values. The most important thing is that the number of reducers will be equal to number of partitions which have been decided by the analysis phase. This becomes a heavy operation because it involves two phases one for finding the ranges of data and other for actually sorting the data which is basically a more of a heavy phase since this actually involves shuffling of the data as obtained. The actual cost of finding the partitions will be very less since it will involve only one reducer and hence the amortized will reduce by a great extent. The order step will involve a lot of data travelling over the network and hence multiple reducers will make the job easier and optimized.

Inverted Index: In recent years, large-scale image retrieval has been shown remarkable potential in real-life applications (Nguyen, B. V., Pham, D., Ngo, T. D., Le, D. D., & Duong, D. A., 2014). This type of job is basically used to list important terms in a dataset and hence is useful for tagging important words in a document so that specific values can be traced out. Most search engines carry out this task for making efficient searches in the web. The mapper gives an output of the desired unique keys attached with values. The partitioner phase is used to determine where data is to be sent to a reducer after processing is done and hence it makes the final processing task of the reducer in a more distributed fashion. Hence finally the reducer receives a collection of distinct row identifiers to link them back to the input keys. Finally the reducers can used with unique delimiters to carry out the operations of separating the key and value pairs. The final output has the values associated with unique IDs from the input file and hence they can be associated with data elements from the file. The performance hence is based upon the unique types of index keys and the cardinality of the keys i.e. the number of values each key is associated with.

CONCLUSION

The golden age of data has arrived. The capacity to handle large data sets problems vary by the server capacity as well as the amount of commodity servers available. By the usage of commodity server the data processing has been scaled out. This was only possible using innovations like map reduce and mpp on the software scale. This mainly involves avoiding system level integrations for the processing tasks and hence making development easier.

REFERENCES

Bijalwan, V., Kumar, V., Kumari, P., & Pascual, J. (2014). KNN based machine learning approach for text and document mining. *International Journal of Database Theory and Application*, *7*(1), 61–70. doi:10.14257/ijdta.2014.7.1.06

Coyle, D. J., Jr., Chang, A., Malkemus, T. R., & Wilson, W. G. (1997). *U.S. Patent No. 5,630,124*. Washington, DC: U.S. Patent and Trademark Office.

Elnikety, S., Dropsho, S., & Pedone, F. (2006, April). Tashkent: Uniting durability with transaction ordering for high-performance scalable database replication. *Operating Systems Review*, *40*(4), 117–130. doi:10.1145/1218063.1217947

Elnikety, S., Pedone, F., & Zwaenepoel, W. (2005, October). Database replication using generalized snapshot isolation. In *Reliable Distributed Systems, 2005. SRDS 2005. 24th IEEE Symposium on* (pp. 73-84). IEEE. 10.1109/RELDIS.2005.14

Eswaran, K. P., Gray, J. N., Lorie, R. A., & Traiger, I. L. (1976). The notions of consistency and predicate locks in a database system. *Communications of the ACM*, *19*(11), 624–633. doi:10.1145/360363.360369

Gupta, A., Agarwal, D., Tan, D., Kulesza, J., Pathak, R., Stefani, S., & Srinivasan, V. (2015, May). Amazon Redshift and the case for simpler data warehouses. In *Proceedings of the 2015 ACM SIGMOD International Conference on Management of Data* (pp. 1917-1923). ACM. 10.1145/2723372.2742795

Kumar, V., Grama, A., Gupta, A., & Karypis, G. (1994). *Introduction to parallel computing: design and analysis of algorithms* (Vol. 400). Redwood City, CA: Benjamin/Cummings.

Metropolis, N. (1986). Massively parallel processing. *Journal of Scientific Computing*, *1*(2), 115–116. doi:10.1007/BF01061388

Nguyen, B. V., Pham, D., Ngo, T. D., Le, D. D., & Duong, D. A. (2014, December). Integrating spatial information into inverted index for large-scale image retrieval. In *Multimedia (ISM), 2014 IEEE International Symposium on* (pp. 102-105). IEEE. doi:10.1007/978-3-319-12024-9_19

Reges, S., & Stepp, M. (2014). *Building Java Programs*. Pearson.

Ristoski, P., Mencía, E. L., & Paulheim, H. (2014, May). A hybrid multi-strategy recommender system using linked open data. In *Semantic Web Evaluation Challenge* (pp. 150–156). Springer International Publishing.

Schatz, M. C. (2009). CloudBurst: Highly sensitive read mapping with MapReduce. *Bioinformatics (Oxford, England)*, *25*(11), 1363–1369. doi:10.1093/bioinformatics/btp236 PMID:19357099

Senthil, G. (2004). *U.S. Patent No. 6,721,869*. Washington, DC: U.S. Patent and Trademark Office.

Taylor, R. C. (2010). An overview of the Hadoop/MapReduce/HBase framework and its current applications in bioinformatics. *BMC Bioinformatics*, 11. PMID:21210976

Tukey, J. W. (1977). *Exploratory data analysis*. Academic Press.

Chapter 10
Image Processing CSE-4019 Distinction Between Fake Note and a Real Note

Shreya Tuli
VIT University, India

Gaurav Sharma
VIT University, India

Nayan Mishr
VIT University, India

ABSTRACT

Big data is a term for data sets that are so large or complex that traditional data processing application software is inadequate to deal with them. Its challenges include capturing data, data storage, data analysis, search, sharing, transfer, visualization, querying, updating, and information privacy. Lately, the term big data tends to refer to the use of predictive analytics, user behavior analytics, or certain other advanced data analytics methods that extract value from data, and seldom to a particular size of data set. In this chapter, the authors distinguish between fake note and a real note and would like to take it to a level where it can be used everywhere. Its data after the detection of the fakeness and the real note can be stored in the database. The data to store will be huge. To overcome this problem, we can go for big data. It will help to store large amounts of data in no time. The difference between real note and fake note is that real note has its thin strip to be more or less continuous while the fake strip has fragmented thin lines in the strip. One could say that the fake note has more than one line in the thin strip while the real note only has one line. Therefore, if we see just one line, it is real, but if we see more than one line, it is fake. In this chapter, the authors use foreign currency.

DOI: 10.4018/978-1-5225-3643-7.ch010

INTRODUCTION

There are 6 different steps in order to distinguish between fake and original currency. In this project, we are using one of the famous segmentation techniques named thresholding. And also, our project involves various process like opening and closing etc…. In this we are using a real note and 2 fake notes and are differentiating between them based on the black strips it has.

Thresholding

It is the simplest method of image segmentation. From a grayscale image, thresholding can be used to create binary images. The simplest thresholding methods replace each pixel in an image with a black pixel if the image intensity is less than some fixed constant T, or a white pixel if the image intensity is greater than that constant. In the example image on the right, this results in the dark tree becoming completely black, and the white snow becoming completely white.

Opening

In mathematical morphology, opening is the dilation of the erosion of a set A by a structuring element B. Together with closing, this serves in computer vision and image processing as a basic workhorse of morphological noise removal. Opening removes small objects from the foreground (usually taken as the bright pixels) of an image, placing them in the background, while closing removes small holes in the foreground, changing small islands of background into foreground. These techniques can also be used to find specific shapes in an image. Opening can be used to find things into which a specific structuring element can fit (edges, corners, ...).

Closing

In mathematical morphology, the closing of a set (binary image) A by a structuring element B is the erosion of the dilation of that set, In image processing, Closing is, together with opening, the basic workhorse of morphological noise removal. Opening removes small objects, while closing removes small holes. This process is to ensure that any disconnected regions that are near larger regions get connected to each other.

PROCEDURE

Let us now see the different steps involved:

Figure 1. Architecture

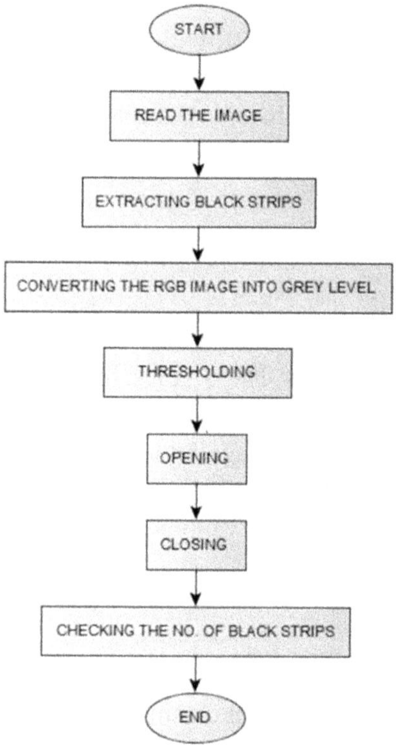

Step 1: Read in the image.

In this step, we are going to read the particular image from a given location, and we also resize the image if it is not in a desired size. Here we are going to directly read our images from desktop using command "imread".

The code for this

```
is: - clear all; close all;
Ireal = imread('C:\Users\Bunny\Desktop\SqbnIm.jpg');
% Real
 Ifake = imread('C:\Users\Bunny\Desktop\2U3DEm.jpg');
% Fake
Ifake2 = imread('C:\Users\Bunny\Desktop\SVJrwaV.jpg');
% Fake #2
% //Resize so that we have the same dimensions as the other
```

Table 1. Literature survey

SNO.	Topic	Abstract
1.	Distinction between the real and the fake one	Paper currency recognition is widely applied in many fields such as bank system and automatic selling-goods system. How to extract highly qualified monetary characteristic vectors from currency image is an important problem needing to be solved now. It is a key process to select original characteristic information from currency image with noises and uneven gray. This article, aiming at the specialties of RENMINGBI (RMB) currency image, puts forward a method using linear transform of image gray to diminish the influence of the background image noises to give prominence to edge information of the image. Then the edge characteristic information image is obtained by edge detection using simple statistics. Finally, by dividing the edge characteristic information image in the width direction into different areas, getting the number of the edge characteristic pixels of different areas as input vectors to neural networks (NN), carrying out sorting recognition by three layer BP NN, paper currency is recognized.
2.	A reliable method for paper currency recognition	Paper currency recognition with good accuracy and high processing speed has great importance for banking system. How to extract high quality monetary features from currency images is a key problem in paper currency recognition. Based on the traditional local binary pattern (LBP) method, an improved LBP algorithm, called block-LBP algorithm, is proposed in this paper for characteristic extraction. The proposed method has advantages of simplicity and high speed. The experimenal results show that this improved method has a high recognition rate, as well as robustness for noise and illumination change.
3.	ANN Based Currency Recognition System using Compressed Gray Scale	Automatic currency note recognition invariably depends on the currency note characteristics of a country and the extraction of features directly affects the recognition ability. Sri Lanka has not been involved in any kind of research or implementation of this kind. The proposed system "SLCRec" comes up with a solution focusing on minimizing false rejection of notes. Sri Lankan currency notes undergo severe changes in image quality in usage. Hence a special linear transformation function is adapted to wipe out noise patterns from backgrounds without affecting the notes' characteristic images and re-appear images of interest.
4.	Feature extraction for paper currency recognition	A new technique for paper currency recognition. In this technique, three characteristics of paper currencies including size, color and texture are used in the recognition. By using image histogram, plenitude of different colors in a paper currency is computed and compared with the one in the reference paper currency. The Markov chain concept has been employed to model texture of the paper currencies as a random process. The method proposed in this paper can be used for recognizing paper currencies from different countries. In this method, using only one intact example of paper currency from each denomination is enough for training the system.

```
images
Ifake2 = imresize (Ifake2, [160 320], 'bilinear'.
```

The output of this step is:

Here the first one is the real image and the rest two are fake ones (Figure 2).

Step 2: Extracting black strips.

Figure 2.

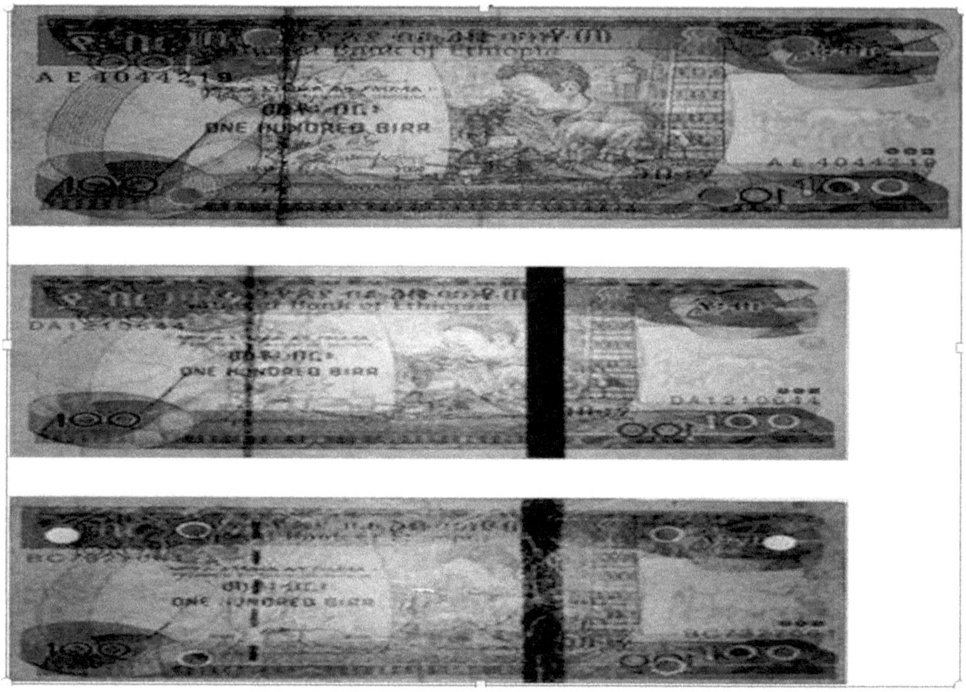

In this step as we initially we have

Discussed that our project is going to work on the principle number of black strips. If black strips are only 1 then it is original. And if we they are not equal to 1 then it is fake image.

The code for this is:

```
BlackStripReal = Ireal (:195:215,:);
 blackStripFake = Ifake(:195:215,:);
 blackStripFake2 = Ifake2(:,195:215,:);
Figure (1);
subplot(1,3,1);
imshow(blackStripReal);
title('Real');
subplot (1,3,2);
imshow(blackStripFake);
title('Fake');
subplot (1,3,3);
 imshow(blackStripFake2);
```

```
title('Fake #2');
In this code, we use command "black strip" for the detection of
black strips in image. Here By looking at all of the notes, the
black strip value hover between the 195th column to the 215th
column.
This is if each image has 320 columns. And subplot here denotes
the 1*3 matrix grid and the particular axes whether that can be
1,2,3 for real, fake,fake2.
```

The output of this step is: (Figure 3).

Step 3: Converting RGB image into gray level and then thresholding

Here in this step we are going to convert our RGB image into gray level image by using the command "rgb2gray". And again, subplot means the same as discussed above. And here we perform thresholding after image is converted into gray scale image. It is done by first converting image into binary by using command "im2bw" and using a threshold value of 30.

Here we used 30 heuristically as this was predominantly the intensity that was for the black strip consisted of.

The code for this step is:

```
BlackStripReal =
rgb2gray(blackStripReal);
```

Figure 3.

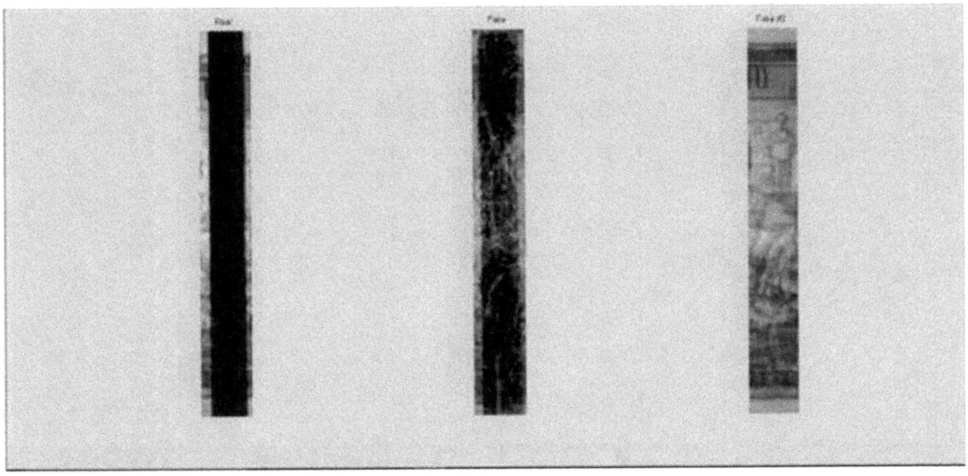

```
blackStripFake = rgb2gray(blackStripFake);
blackStripFake2 = rgb2gray(blackStripFake2);
Figure (2);
subplot (1,3,1);
imshow (blackStripReal);
title('Real');
subplot (1,3,2);
 imshow(blackStripFake);
 title('Fake');
subplot (1,3,3);
imshow(blackStripFake2);
 title ('Fake #2');
```

The output for this is: (Figure 4).

For thresholding: -

```
blackStripRealBW = ~im2bw (blackStripReal, 30/255);
blackStripFakeBW = ~im2bw (blackStripFake, 30/255);
blackStripFake2BW = ~im2bw (blackStripFake2, 30/255);
figure (3);
subplot (1,3,1);
imshow(blackStripRealBW);
title('Real');
```

Figure 4.

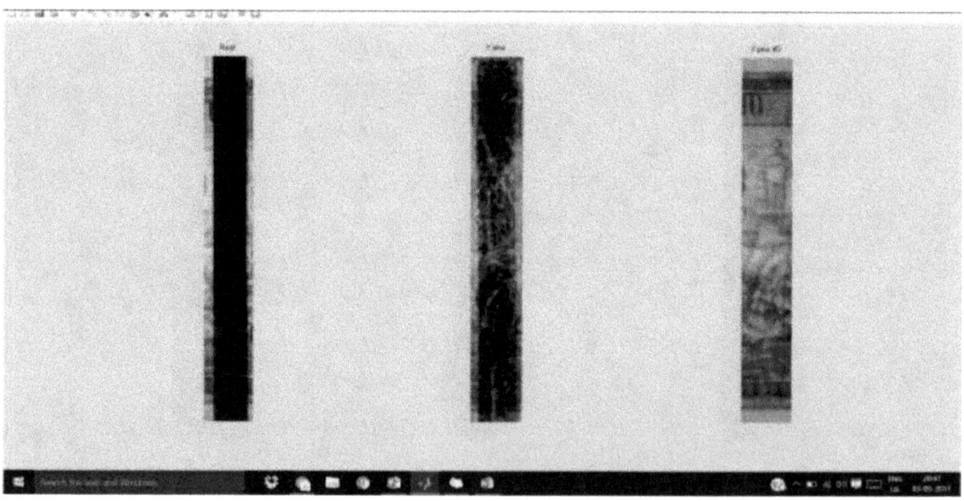

```
subplot (1,3,2);
 imshow(blackStripFakeBW);
 title('Fake');
subplot (1,3,3);
 imshow(blackStripFake2BW);
title ('Fake #2');
Here again we use subplot for the same reason as discussed
above:
```

The output for this: (Figure 5).

Step 4: Opening.

In this step we are doing area opening of the image like by specifying a larger area of about 100 to ensure that the image will get rid of any spurious noisy and isolated pixels. You'll notice that for each of the images, there are some noisy pixels on the edges. We perform this opening operation using the command "bwareaopen". This function removes pixel areas in a black and white image that have less than a certain area.

The code for this is:

```
areaopenReal = bwareaopen (blackStripRealBW, 100);
subplot(1,3,1);
```

Figure 5.

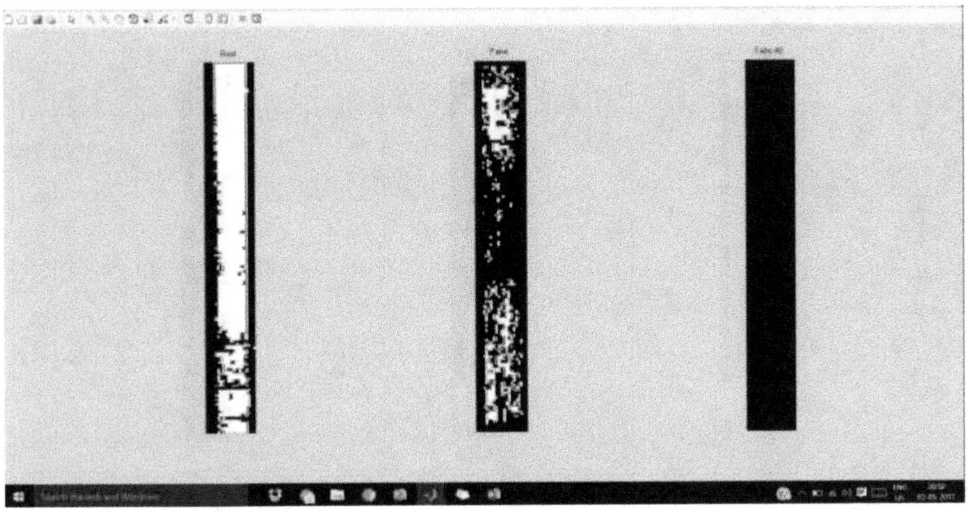

```
imshow(areaopenReal);
title(‘Real’);
subplot (1,3,2);
areaopenFake = bwareaopen (blackStripFakeBW, 100);
imshow(areaopenFake);
title(‘Fake’);
subplot (1,3,3);
areaopenFake2 =  bwareaopen (blackStripFake2BW, 100);
imshow(areaopenFake2);
title (‘Fake #2’);
```

Here also again we use subplot for the same reason discussed above.
The output for this is: (Figure 6).

Step 5: Post Process.

This is nothing but closing process. Here in this process we use a square structuring element of 5 x 5 to ensure that any disconnected regions that are near larger regions get connected to each other. For this process to be done we use command “imclose”.

The code for this process is:

```
se = strel (‘square’, 5);
BWImageCloseReal = imclose (areaopenReal, se);
```

Figure 6.

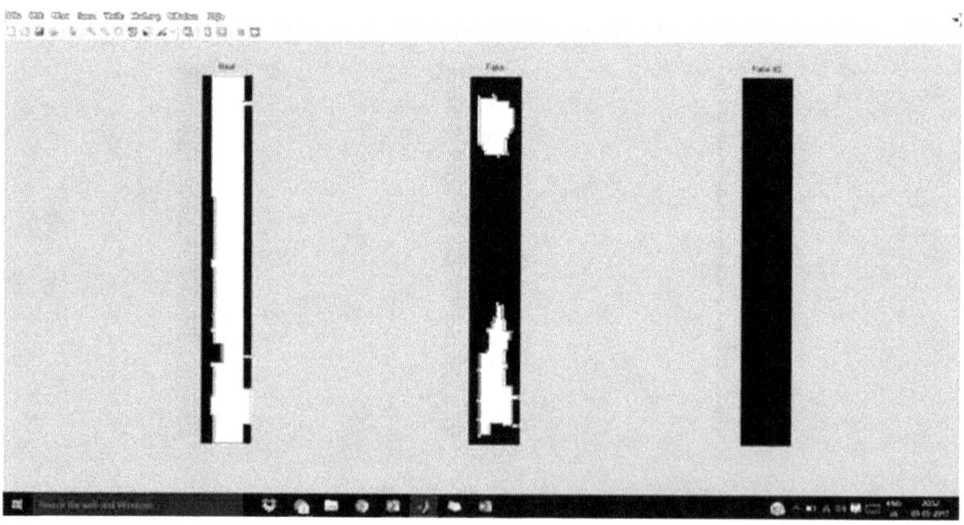

```
BWImageCloseFake = imclose (areaopenFake, se);
BWImageCloseFake2 = imclose (areaopenFake2, se);
Figure(5);
Subplot (1,3,1);
imshow(BWImageCloseReal);
title('Real');
Subplot (1,3,2);
imshow(BWImageCloseFake);
title('Fake');
Subplot (1,3,3);
imshow(BWImageCloseFake2);
 title ('Fake #2');
```

Here again we use subplot for the same reason as discussed above.
The output for this is: (Figure 7).

Step 6: Counting the total number of objects in this strip.

The last step is to simply count the number of black lines in each image. If there is just 1, this denotes that the bank note is real, while if there is more than 1, this denotes that the bank note is fake. For this we use the command "bwlabel" and use the second parameter to count how many objects are there.

Figure 7.

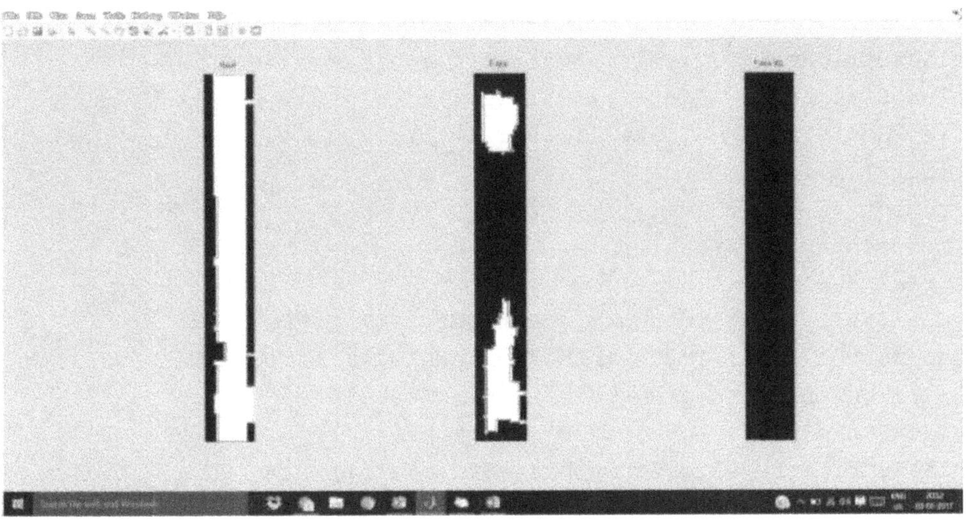

The code for this is:

```
[~,countReal] = bwlabel(BWImageCloseReal);
[~,countFake] = bwlabel(BWImageCloseFake);
 [~,countFake2] = bwlabel(BWImageCloseFake2);
disp (['The total number of black lines for the real note is: '
num2str(countReal)]);
Disp (['The total number of black lines for the fake note is: '
num2str(countFake)]);
Disp (['The total number of black lines for the second fake
note is: ' num2str(countFake2)]);
Finally, we use "disp" to display the output:
The total number of black lines for the real note is: 1
The total number of black lines for the fake note is: 2
The total number of black lines for the second fake note is: 0
```

FULL MATLAB CODE

```
clear all;
close all;
Ireal = imread('C:\Users\Bunny\Desktop\SqbnIm.jpg'); % Real
Ifake = imread('C:\Users\Bunny\Desktop\2U3DEm.jpg'); % Fake
Ifake2 = imread('C:\Users\Bunny\Desktop\SVJrwaV.jpg'); % Fake
#2
% //Resize so that we have the same dimensions as the other
images
Ifake2 = imresize (Ifake2, [160 320], 'bilinear');
%% //Extract the black strips for each image
blackStripReal = Ireal(:,195:215,:);
blackStripFake = Ifake(:,195:215,:);
blackStripFake2 = Ifake2(:,195:215,:);
Figure (1);
subplot (1,3,1);
imshow(blackStripReal);
title('Real');
subplot (1,3,2); imshow(blackStripFake); title('Fake');
subplot (1,3,3); imshow(blackStripFake2); title ('Fake #2');
%% //Convert into grayscale then threshold
blackStripReal = rgb2gray(blackStripReal);
blackStripFake = rgb2gray(blackStripFake);
blackStripFake2 = rgb2gray(blackStripFake2);
Figure (2);
```

```
subplot (1,3,1);
imshow (blackStripReal);
title ('Real');
subplot (1,3,2);
imshow (blackStripFake);
title ('Fake');
subplot (1,3,3);
imshow (blackStripFake2);
title ('Fake #2');
%% //Threshold using about intensity 30
blackStripRealBW = ~im2bw (blackStripReal, 30/255);
blackStripFakeBW = ~im2bw (blackStripFake, 30/255);
blackStripFake2BW = ~im2bw (blackStripFake2, 30/255);
figure (3);
subplot (1,3,1);
imshow (blackStripRealBW);
title ('Real');
subplot (1,3,2);
imshow (blackStripFakeBW);
title ('Fake');
subplot (1,3,3);
imshow (blackStripFake2BW);
title ('Fake #2');
%% //Area open the image
Figure (4);
areaopenReal = bwareaopen (blackStripRealBW, 100);
subplot (1,3,1);
imshow (areaopenReal);
title ('Real');
subplot (1,3,2);
areaopenFake = bwareaopen (blackStripFakeBW, 100);
imshow (areaopenFake);
title ('Fake');
subplot (1,3,3);
areaopenFake2 = bwareaopen (blackStripFake2BW, 100);
imshow (areaopenFake2);
title ('Fake #2');
%% //Post-process
se = strel ('square', 5);
BWImageCloseReal = imclose (areaopenReal, se);
```

```
BWImageCloseFake = imclose (areaopenFake, se);
BWImageCloseFake2 = imclose (areaopenFake2, se);
Figure (5);
Subplot (1,3,1);
Imshow (BWImageCloseReal);
Title ('Real');
Subplot (1,3,2);
Imshow (BWImageCloseFake);
Title ('Fake');
Subplot (1,3,3);
Imshow (BWImageCloseFake2);
Title ('Fake #2');
%% //Count the total number of objects in this strip
[~, countReal] = bwlabel (BWImageCloseReal);
[~, countFake] = bwlabel (BWImageCloseFake);
[~, countFake2] = bwlabel (BWImageCloseFake2);
Disp (['The total number of black lines for the real note is: '
num2str(countReal)]);
Disp (['The total number of black lines for the fake note is:
'num2str(countFake)]);
disp(['The total number of black lines for the second fake note
is: '
num2str(countFake2)]);
RESULTS
```

Figure 8.

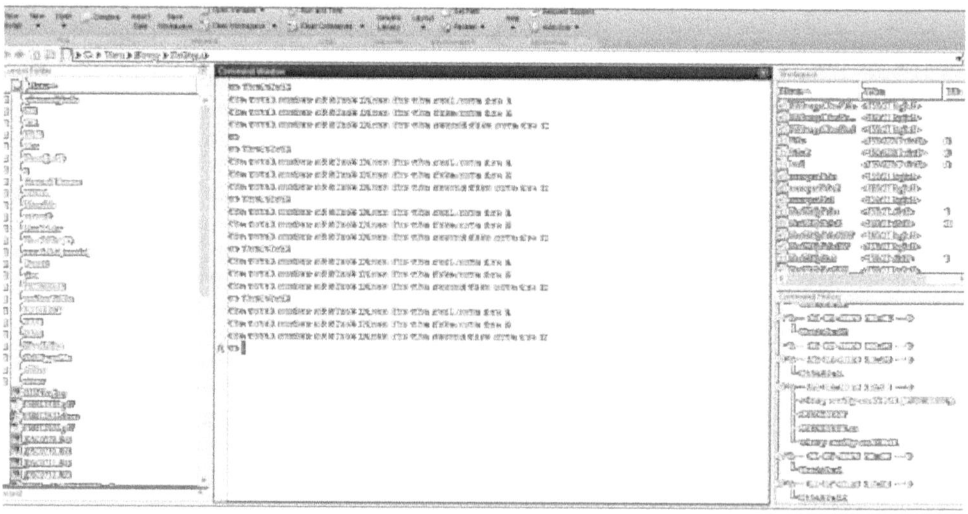

As seen above these are the results that we obtain while we run this process.

CONCLUSION

The conclusion is that we are using various image processing techniques to determine whether a note is fake or real. We are telling this based on number of black strips that are present on the image. We have used the processes like thresholding, opening, closing etc. Here, I would like to conclude the big data helps us to overcome the problem of storing a large amount of data instantly and with a great efficiency.

Big Data depends on the following 3V's:

- **Volume:** Big data doesn't sample; it just observes and tracks what happens.
- **Velocity:** Big data is often available in real-time.

Figure 9.

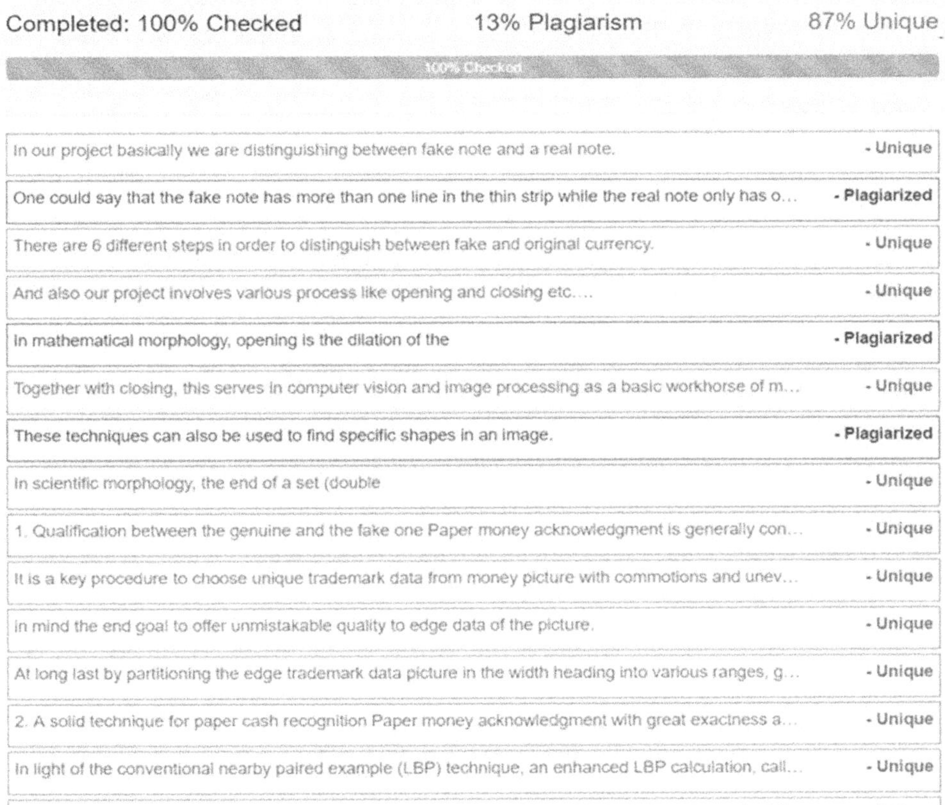

- **Variety:** Big data draws from text, images, audio, video; plus it completes missing pieces through a fusion.

FUTURE SCOPE

This project has a good future scope. Now-a-days most of the people print these fake notes which are illegal. So, this process can be used by the bank people to detect which is the real and fake currency which is very helpful.

Bid Data in the coming time will play a major role in storing a large amount of data for storing the information of how many fake and how many real notes were detected in the set.

REFERENCES

Gunaratna, Kodikara, & Premaratne. (2008). ANN based currency recognition system using compressed gray scale and application for Sri Lankan currency notes-SLCRec. *Proceedings of World Academy of Science, Engineering and Technology, 35*, 235-240.

Guo, Zhao, & Cai. (2010). A reliable method for paper currency recognition based on LBP. In *Network Infrastructure and Digital Content, 2010 2nd IEEE International Conference on*. IEEE.

Hassanpour, Yaseri, & Ardeshiri. (2007). Feature extraction for paper currency recognition. In *Signal Processing and Its Applications,* 2007. *ISSPA 2007. 9th International Symposium on*. IEEE.

Zhang, E.-H. (2003). Research on paper currency recognition by neural networks. In *Machine Learning and Cybernetics, 2003 International Conference on* (vol. 4). IEEE.

Related References

To continue our tradition of advancing knowledge management and discovery research, we have compiled a list of recommended IGI Global readings. These references will provide additional information and guidance to further enrich your knowledge and assist you with your own research and future publications.

Abril, R. M. (2011). The quality attribution in data, information and knowledge. In D. Schwartz & D. Te'eni (Eds.), *Encyclopedia of knowledge management* (2nd ed.; pp. 1343–1354). Hershey, PA: IGI Global. doi:10.4018/978-1-59904-931-1.ch129

Abufardeh, S. (2013). KM and global software engineering (GSE). In S. Saeed & I. Alsmadi (Eds.), *Knowledge-based processes in software development* (pp. 12–34). Hershey, PA: IGI Global. doi:10.4018/978-1-4666-4229-4.ch002

Aggestam, L., Backlund, P., & Persson, A. (2010). Supporting knowledge evaluation to increase quality in electronic knowledge repositories. *International Journal of Knowledge Management, 6*(1), 23–43. doi:10.4018/jkm.2010103002

Aggestam, L., Backlund, P., & Persson, A. (2012). Supporting knowledge evaluation to increase quality in electronic knowledge repositories. In M. Jennex (Ed.), *Conceptual models and outcomes of advancing knowledge management: New technologies* (pp. 24–44). Hershey, PA: IGI Global. doi:10.4018/978-1-4666-0035-5.ch002

Aiken, P., Gillenson, M., Zhang, X., & Rafner, D. (2011). Data management and data administration: Assessing 25 years of practice. *Journal of Database Management, 22*(3), 24–45. doi:10.4018/jdm.2011070102

Akabawi, S., & Hodeeb, H. (2013). Implementing business intelligence in the dynamic beverages sales and distribution environment. In M. Khosrow-Pour (Ed.), *Cases on performance measurement and productivity improvement: Technology integration and maturity* (pp. 194–221). Hershey, PA: IGI Global. doi:10.4018/978-1-4666-2618-8.ch010

Al-Busaidi, K. A. (2011). A social and technical investigation of knowledge utilization from a repository knowledge management system. In M. Al-Shammari (Ed.), *Knowledge management in emerging economies: Social, organizational and cultural implementation* (pp. 122–139). Hershey, PA: IGI Global. doi:10.4018/978-1-61692-886-5.ch007

Al-Busaidi, K. A. (2012). The impact of supporting organizational knowledge management through a corporate portal on employees and business processes. In M. Jennex (Ed.), *Conceptual models and outcomes of advancing knowledge management: New technologies* (pp. 208–229). Hershey, PA: IGI Global. doi:10.4018/978-1-4666-0035-5.ch011

Alaraifi, A., Molla, A., & Deng, H. (2013). An empirical analysis of antecedents to the assimilation of sensor information systems in data centers. *International Journal of Information Technologies and Systems Approach*, *6*(1), 57–77. doi:10.4018/jitsa.2013010104

Alguezaui, S., & Filieri, R. (2010). Social capital: Knowledge and technological innovation. In P. López Sáez, G. Castro, J. Navas López, & M. Delgado Verde (Eds.), *Intellectual capital and technological innovation: Knowledge-based theory and practice* (pp. 271–296). Hershey, PA: IGI Global. doi:10.4018/978-1-61520-875-3.ch013

Alhashem, A., & Shaqrah, A. A. (2012). Exploring the relationship between organizational memory and business innovation. *International Journal of Knowledge-Based Organizations*, *2*(3), 32–46. doi:10.4018/ijkbo.2012070102

Allan, M. B., Korolis, A. A., & Griffith, T. L. (2011). Reaching for the moon: Expanding transactive memory's reach with wikis and tagging. In M. Jennex (Ed.), *Global aspects and cultural perspectives on knowledge management: Emerging dimensions* (pp. 144–156). Hershey, PA: IGI Global. doi:10.4018/978-1-60960-555-1.ch010

Alsmadi, I., & Alda, S. (2013). Knowledge management and semantic web services. In S. Saeed & I. Alsmadi (Eds.), *Knowledge-based processes in software development* (pp. 35–48). Hershey, PA: IGI Global. doi:10.4018/978-1-4666-4229-4.ch003

Alstete, J. W., & Meyer, J. P. (2011). Expanding the model of competitive business strategy for knowledge-based organizations. *International Journal of Knowledge-Based Organizations, 1*(4), 16–31. doi:10.4018/ijkbo.2011100102

Alstete, J. W., & Meyer, J. P. (2013). Expanding the model of competitive business strategy for knowledge-based organizations. In J. Wang (Ed.), *Intelligence methods and systems advancements for knowledge-based business* (pp. 132–148). Hershey, PA: IGI Global. doi:10.4018/978-1-4666-1873-2.ch008

Alves da Silva, N. S., Alvarez, I. M., & Rogerson, S. (2011). Glocality, diversity and ethics of distributed knowledge in higher education. In G. Morais da Costa (Ed.), *Ethical issues and social dilemmas in knowledge management: Organizational innovation* (pp. 131–159). Hershey, PA: IGI Global. doi:10.4018/978-1-61520-873-9.ch009

Anantatmula, V. S., & Kanungo, S. (2011). Strategies for successful implementation of KM in a university setting. In M. Jennex & S. Smolnik (Eds.), *Strategies for knowledge management success: Exploring organizational efficacy* (pp. 262–276). Hershey, PA: IGI Global. doi:10.4018/978-1-60566-709-6.ch014

Andreu, R., & Sieber, S. (2011). External and internal knowledge in organizations. In D. Schwartz & D. Te'eni (Eds.), *Encyclopedia of knowledge management* (2nd ed.; pp. 298–307). Hershey, PA: IGI Global. doi:10.4018/978-1-59904-931-1.ch029

Andriessen, D. (2011). Metaphor use in knowledge management. In D. Schwartz & D. Te'eni (Eds.), *Encyclopedia of knowledge management* (2nd ed.; pp. 1118–1124). Hershey, PA: IGI Global. doi:10.4018/978-1-59904-931-1.ch107

Angelopoulos, S., Kitsios, F., & Moustakis, V. (2012). Transformation of management in the public sector: Exploring the strategic frameworks of e-government. In T. Papadopoulos & P. Kanellis (Eds.), *Public sector reform using information technologies: Transforming policy into practice* (pp. 44–58). Hershey, PA: IGI Global. doi:10.4018/978-1-60960-839-2.ch003

Anselma, L., Bottrighi, A., Molino, G., Montani, S., Terenziani, P., & Torchio, M. (2013). Supporting knowledge-based decision making in the medical context: The GLARE approach. In J. Wang (Ed.), *Intelligence methods and systems advancements for knowledge-based business* (pp. 24–42). Hershey, PA: IGI Global. doi:10.4018/978-1-4666-1873-2.ch002

Arh, T., Dimovski, V., & Blažic, B. J. (2011). ICT and web 2.0 technologies as a determinant of business performance. In M. Al-Mutairi & L. Mohammed (Eds.), *Cases on ICT utilization, practice and solutions: Tools for managing day-to-day issues* (pp. 59–77). Hershey, PA: IGI Global. doi:10.4018/978-1-60960-015-0.ch005

Ariely, G. (2011). Operational knowledge management in the military. In D. Schwartz & D. Te'eni (Eds.), *Encyclopedia of knowledge management* (2nd ed.; pp. 1250–1260). Hershey, PA: IGI Global. doi:10.4018/978-1-59904-931-1.ch119

Assefa, T., Garfield, M., & Meshesha, M. (2014). Enabling factors for knowledge sharing among employees in the workplace. In Y. Al-Bastaki & A. Shajera (Eds.), *Building a competitive public sector with knowledge management strategy* (pp. 246–271). Hershey, PA: IGI Global. doi:10.4018/978-1-4666-4434-2.ch011

Assudani, R. H. (2011). Negotiating knowledge gaps in dispersed knowledge work. *International Journal of Knowledge-Based Organizations, 1*(3), 1–21. doi:10.4018/ijkbo.2011070101

Assudani, R. H. (2013). Negotiating knowledge gaps in dispersed knowledge work. In J. Wang (Ed.), *Intelligence methods and systems advancements for knowledge-based business* (pp. 75–96). Hershey, PA: IGI Global. doi:10.4018/978-1-4666-1873-2.ch005

Atkins, R. (2011). Supply chain knowledge integration in emerging economies. In M. Al-Shammari (Ed.), *Knowledge management in emerging economies: Social, organizational and cultural implementation* (pp. 104–121). Hershey, PA: IGI Global. doi:10.4018/978-1-61692-886-5.ch006

Aung, Z., & Nyunt, K. K. (2014). Constructive knowledge management model and information retrieval methods for software engineering. In *Software design and development: Concepts, methodologies, tools, and applications* (pp. 253–269). Hershey, PA: IGI Global. doi:10.4018/978-1-4666-4301-7.ch014

Aziz, M. W., Mohamad, R., & Jawawi, D. N. (2013). Ontology-based service description, discovery, and matching in distributed embedded real-time systems. In M. Nazir Ahmad, R. Colomb, & M. Abdullah (Eds.), *Ontology-based applications for enterprise systems and knowledge management* (pp. 178–190). Hershey, PA: IGI Global. doi:10.4018/978-1-4666-1993-7.ch010

Badia, A. (2011). Knowledge management and intelligence work: A promising combination. In D. Schwartz & D. Te'eni (Eds.), *Encyclopedia of knowledge management* (2nd ed.; pp. 612–623). Hershey, PA: IGI Global. doi:10.4018/978-1-59904-931-1.ch059

Badr, K. B., & Ahmad, M. N. (2013). Managing lessons learned: A comparative study of lessons learned systems. In M. Nazir Ahmad, R. Colomb, & M. Abdullah (Eds.), *Ontology-based applications for enterprise systems and knowledge management* (pp. 224–245). Hershey, PA: IGI Global. doi:10.4018/978-1-4666-1993-7.ch013

Bakshi, K. (2014). Technologies for big data. In W. Hu & N. Kaabouch (Eds.), *Big data management, technologies, and applications* (pp. 1–22). Hershey, PA: IGI Global. doi:10.4018/978-1-4666-4699-5.ch001

Ballou, D. P., Belardo, S., & Pazer, H. L. (2012). A project staffing model to enhance the effectiveness of knowledge transfer in the requirements planning phase for multi-project environments. In M. Jennex (Ed.), *Conceptual models and outcomes of advancing knowledge management: New technologies* (pp. 77–98). Hershey, PA: IGI Global. doi:10.4018/978-1-4666-0035-5.ch005

Baporikar, N. (2011). Knowledge management and entrepreneurship cases in India. In M. Al-Shammari (Ed.), *Knowledge management in emerging economies: Social, organizational and cultural implementation* (pp. 325–346). Hershey, PA: IGI Global. doi:10.4018/978-1-61692-886-5.ch020

Baporikar, N. (2014). Knowledge management initiatives in Indian public sector. In Y. Al-Bastaki & A. Shajera (Eds.), *Building a competitive public sector with knowledge management strategy* (pp. 53–89). Hershey, PA: IGI Global. doi:10.4018/978-1-4666-4434-2.ch002

Baporikar, N. (2014). Organizational barriers and facilitators in embedding knowledge strategy. In M. Chilton & J. Bloodgood (Eds.), *Knowledge management and competitive advantage: Issues and potential solutions* (pp. 149–173). Hershey, PA: IGI Global. doi:10.4018/978-1-4666-4679-7.ch009

Barioni, M. C., Kaster, D. D., Razente, H. L., Traina, A. J., & Júnior, C. T. (2011). Querying multimedia data by similarity in relational DBMS. In L. Yan & Z. Ma (Eds.), *Advanced database query systems: Techniques, applications and technologies* (pp. 323–359). Hershey, PA: IGI Global. doi:10.4018/978-1-60960-475-2.ch014

Baroni de Carvalho, R., & Tavares Ferreira, M. A. (2011). Knowledge management software. In D. Schwartz & D. Te'eni (Eds.), *Encyclopedia of knowledge management* (2nd ed.; pp. 738–749). Hershey, PA: IGI Global. doi:10.4018/978-1-59904-931-1.ch072

Barroso, A. C., Ricciardi, R. I., & Junior, J. A. (2012). Web 2.0 and project management: Reviewing the change path and discussing a few cases. In I. Boughzala & A. Dudezert (Eds.), *Knowledge management 2.0: Organizational models and enterprise strategies* (pp. 164–189). Hershey, PA: IGI Global. doi:10.4018/978-1-61350-195-5.ch009

Baskaran, V., Naguib, R., Guergachi, A., Bali, R., & Arochen, H. (2011). Does knowledge management really work? A case study in the breast cancer screening domain. In A. Eardley & L. Uden (Eds.), *Innovative knowledge management: Concepts for organizational creativity and collaborative design* (pp. 177–189). Hershey, PA: IGI Global. doi:10.4018/978-1-60566-701-0.ch010

Bebensee, T., Helms, R., & Spruit, M. (2012). Exploring the impact of web 2.0 on knowledge management. In I. Boughzala & A. Dudezert (Eds.), *Knowledge management 2.0: Organizational models and enterprise strategies* (pp. 17–43). Hershey, PA: IGI Global. doi:10.4018/978-1-61350-195-5.ch002

Becerra-Fernandez, I., & Sabherwal, R. (2011). The role of information and communication technologies in knowledge management: A classification of knowledge management systems. In D. Schwartz & D. Te'eni (Eds.), *Encyclopedia of knowledge management* (2nd ed.; pp. 1410–1418). Hershey, PA: IGI Global. doi:10.4018/978-1-59904-931-1.ch134

Benbya, H. (2013). Valuing knowledge-based initiatives: What we know and what we don't know. In M. Jennex (Ed.), *Dynamic models for knowledge-driven organizations* (pp. 1–15). Hershey, PA: IGI Global. doi:10.4018/978-1-4666-2485-6.ch001

Berends, H., van der Bij, H., & Weggeman, M. (2011). Knowledge integration. In D. Schwartz & D. Te'eni (Eds.), *Encyclopedia of knowledge management* (2nd ed.; pp. 581–590). Hershey, PA: IGI Global. doi:10.4018/978-1-59904-931-1.ch056

Berger, H., & Beynon-Davies, P. (2011). Knowledge-based diffusion in practice: A case study experience. In A. Eardley & L. Uden (Eds.), *Innovative knowledge management: Concepts for organizational creativity and collaborative design* (pp. 40–55). Hershey, PA: IGI Global. doi:10.4018/978-1-60566-701-0.ch003

Berio, G., Di Leva, A., Harzallah, M., & Sacco, G. M. (2012). Competence management over social networks through dynamic taxonomies. In I. Boughzala & A. Dudezert (Eds.), *Knowledge management 2.0: Organizational models and enterprise strategies* (pp. 103–120). Hershey, PA: IGI Global. doi:10.4018/978-1-61350-195-5.ch006

Bhatt, S., Chaudhary, S., & Bhise, M. (2013). Migration of data between cloud and non-cloud datastores. In A. Ionita, M. Litoiu, & G. Lewis (Eds.), *Migrating legacy applications: Challenges in service oriented architecture and cloud computing environments* (pp. 206–225). Hershey, PA: IGI Global. doi:10.4018/978-1-4666-2488-7.ch009

Bloodgood, J. M., Chilton, M. A., & Bloodgood, T. C. (2014). The effect of knowledge transfer motivation, receiver capability, and motivation on organizational performance. In M. Chilton & J. Bloodgood (Eds.), *Knowledge management and competitive advantage: Issues and potential solutions* (pp. 232–242). Hershey, PA: IGI Global. doi:10.4018/978-1-4666-4679-7.ch013

Boersma, K., & Kingma, S. (2011). Organizational learning facilitation with intranet (2.0): A socio-cultural approach. In D. Schwartz & D. Te'eni (Eds.), *Encyclopedia of knowledge management* (2nd ed., pp. 1280–1289). Hershey, PA: IGI Global. doi:10.4018/978-1-59904-931-1.ch122

Bond, P. L. (2010). Toward a living systems framework for unifying technology and knowledge management, organizational, cultural and economic change. In D. Harorimana (Ed.), *Cultural implications of knowledge sharing, management and transfer: Identifying competitive advantage* (pp. 108–132). Hershey, PA: IGI Global. doi:10.4018/978-1-60566-790-4.ch006

Bordogna, G., Bucci, F., Carrara, P., Pepe, M., & Rampini, A. (2011). Flexible querying of imperfect temporal metadata in spatial data infrastructures. In L. Yan & Z. Ma (Eds.), *Advanced database query systems: Techniques, applications and technologies* (pp. 140–159). Hershey, PA: IGI Global. doi:10.4018/978-1-60960-475-2.ch006

Boughzala, I. (2012). Collaboration 2.0 through the new organization (2.0) transformation. In I. Boughzala & A. Dudezert (Eds.), *Knowledge management 2.0: Organizational models and enterprise strategies* (pp. 1–16). Hershey, PA: IGI Global. doi:10.4018/978-1-61350-195-5.ch001

Bratianu, C. (2011). A new perspective of the intellectual capital dynamics in organizations. In B. Vallejo-Alonso, A. Rodriguez-Castellanos, & G. Arregui-Ayastuy (Eds.), *Identifying, measuring, and valuing knowledge-based intangible assets: New perspectives* (pp. 1–21). Hershey, PA: IGI Global. doi:10.4018/978-1-60960-054-9.ch001

Bratianu, C. (2011). Universities as knowledge-intensive learning organizations. In A. Eardley & L. Uden (Eds.), *Innovative knowledge management: Concepts for organizational creativity and collaborative design* (pp. 1–17). Hershey, PA: IGI Global. doi:10.4018/978-1-60566-701-0.ch001

Breu, K., Ward, J., & Murray, P. (2000). Success factors in leveraging the corporate information and knowledge resource through intranets. In Y. Malhotra (Ed.), *Knowledge management and virtual organizations* (pp. 306–320). Hershey, PA: IGI Global. doi:10.4018/978-1-930708-65-5.ch016

Briones-Peñalver, A., & Poças-Rascão, J. (2014). Information technologies (ICT), network organizations, and information systems for business cooperation: A focus on organization and strategic knowledge management. In G. Jamil, A. Malheiro, & F. Ribeiro (Eds.), *Rethinking the conceptual base for new practical applications in information value and quality* (pp. 324–348). Hershey, PA: IGI Global. doi:10.4018/978-1-4666-4562-2.ch015

Brock, J. K., & Zhou, Y. J. (2011). MNE knowledge management across borders and ICT. In D. Schwartz & D. Te'eni (Eds.), *Encyclopedia of knowledge management* (2nd ed.; pp. 1136–1148). Hershey, PA: IGI Global. doi:10.4018/978-1-59904-931-1.ch109

Brunet-Thornton, R., & Bureš, V. (2011). Meeting Czech knowledge management challenges head-on: KM-Be.At-It. In M. Al-Shammari (Ed.), *Knowledge management in emerging economies: Social, organizational and cultural implementation* (pp. 20–46). Hershey, PA: IGI Global. doi:10.4018/978-1-61692-886-5.ch002

Bucher, T., & Dinter, B. (2012). Situational method engineering to support process-oriented information logistics: Identification of development situations. *Journal of Database Management, 23*(1), 31–48. doi:10.4018/jdm.2012010102

Burstein, F., & Linger, H. (2011). Task-based knowledge management approach. In D. Schwartz & D. Te'eni (Eds.), *Encyclopedia of knowledge management* (2nd ed.; pp. 1479–1489). Hershey, PA: IGI Global. doi:10.4018/978-1-59904-931-1.ch141

Butler, T. (2011). Anti-foundational knowledge management. In D. Schwartz & D. Te'eni (Eds.), *Encyclopedia of knowledge management* (2nd ed.; pp. 1–11). Hershey, PA: IGI Global. doi:10.4018/978-1-59904-931-1.ch001

Butler, T., & Murphy, C. (2011). Work and knowledge. In D. Schwartz & D. Te'eni (Eds.), *Encyclopedia of knowledge management* (2nd ed.; pp. 1556–1566). Hershey, PA: IGI Global. doi:10.4018/978-1-59904-931-1.ch148

Cabrilo, S., & Grubic-Nesic, L. (2013). The role of creativity, innovation, and invention in knowledge management. In S. Buckley & M. Jakovljevic (Eds.), *Knowledge management innovations for interdisciplinary education: Organizational applications* (pp. 207–232). Hershey, PA: IGI Global. doi:10.4018/978-1-4666-1969-2.ch011

Cabrita, M. D., Machado, V. C., & Grilo, A. (2010). Intellectual capital: How knowledge creates value. In E. O'Brien, S. Clifford, & M. Southern (Eds.), *Knowledge management for process, organizational and marketing innovation: Tools and methods* (pp. 237–252). Hershey, PA: IGI Global. doi:10.4018/978-1-61520-829-6.ch015

Cagliero, L., & Fiori, A. (2013). Knowledge discovery from online communities. In *Data mining: Concepts, methodologies, tools, and applications* (pp. 1230–1252). Hershey, PA: IGI Global. doi:10.4018/978-1-4666-2455-9.ch063

Camisón-Zornoza, C., & Boronat-Navarro, M. (2010). Linking exploration and exploitation capabilities with the process of knowledge development and with organizational facilitators. In M. Russ (Ed.), *Knowledge management strategies for business development* (pp. 159–179). Hershey, PA: IGI Global. doi:10.4018/978-1-60566-348-7.ch008

Carneiro, A. (2010). Change knowledge management: Transforming a ghost community into a real asset. In E. O'Brien, S. Clifford, & M. Southern (Eds.), *Knowledge management for process, organizational and marketing innovation: Tools and methods* (pp. 120–132). Hershey, PA: IGI Global. doi:10.4018/978-1-61520-829-6.ch007

Carrillo, F. J. (2010). Knowledge-based value generation. In K. Metaxiotis, F. Carrillo, & T. Yigitcanlar (Eds.), *Knowledge-based development for cities and societies: Integrated multi-level approaches* (pp. 1–16). Hershey, PA: IGI Global. doi:10.4018/978-1-61520-721-3.ch001

Cartelli, A. (2012). Frameworks for the benchmarking of digital and knowledge management best practice in SME and organizations. In A. Cartelli (Ed.), *Current trends and future practices for digital literacy and competence* (pp. 166–175). Hershey, PA: IGI Global. doi:10.4018/978-1-4666-0903-7.ch015

Castellano, G., Fanelli, A. M., & Torsello, M. A. (2010). Soft computing techniques in content-based multimedia information retrieval. In K. Anbumani & R. Nedunchezhian (Eds.), *Soft computing applications for database technologies: Techniques and issues* (pp. 170–192). Hershey, PA: IGI Global. doi:10.4018/978-1-60566-814-7.ch010

Chalkiti, K., & Carson, D. (2010). Knowledge cultures, competitive advantage and staff turnover in hospitality in Australia's northern territory. In D. Harorimana (Ed.), *Cultural implications of knowledge sharing, management and transfer: Identifying competitive advantage* (pp. 203–229). Hershey, PA: IGI Global. doi:10.4018/978-1-60566-790-4.ch010

Chang, W., & Li, S. (2011). Deploying knowledge management in R&D workspaces. In A. Eardley & L. Uden (Eds.), *Innovative knowledge management: Concepts for organizational creativity and collaborative design* (pp. 56–76). Hershey, PA: IGI Global. doi:10.4018/978-1-60566-701-0.ch004

Chawla, D., & Joshi, H. (2013). Impact of knowledge management dimensions on learning organization: Comparison across business excellence awarded and non-awarded indian organizations. In M. Jennex (Ed.), *Dynamic models for knowledge-driven organizations* (pp. 145–162). Hershey, PA: IGI Global. doi:10.4018/978-1-4666-2485-6.ch008

Chen, E. (2012). Web 2.0 social networking technologies and strategies for knowledge management. In I. Boughzala & A. Dudezert (Eds.), *Knowledge management 2.0: Organizational models and enterprise strategies* (pp. 84–102). Hershey, PA: IGI Global. doi:10.4018/978-1-61350-195-5.ch005

Chihara, K., & Nakamori, Y. (2013). Clarification of abilities and qualities of knowledge coordinators: The case of regional revitalization projects. In G. Yang (Ed.), *Multidisciplinary studies in knowledge and systems science* (pp. 1–17). Hershey, PA: IGI Global. doi:10.4018/978-1-4666-3998-0.ch001

Christidis, K., Papailiou, N., Apostolou, D., & Mentzas, G. (2011). Semantic interfaces for personal and social knowledge work. *International Journal of Knowledge-Based Organizations, 1*(1), 61–77. doi:10.4018/ijkbo.2011010104

Christidis, K., Papailiou, N., Apostolou, D., & Mentzas, G. (2013). Semantic interfaces for personal and social knowledge work. In J. Wang (Ed.), *Intelligence methods and systems advancements for knowledge-based business* (pp. 213–230). Hershey, PA: IGI Global. doi:10.4018/978-1-4666-1873-2.ch012

Chua, C. E., Storey, V. C., & Chiang, R. H. (2012). Knowledge representation: A conceptual modeling approach. *Journal of Database Management, 23*(1), 1–30. doi:10.4018/jdm.2012010101

Clinton, M. S., Merritt, K. L., & Murray, S. R. (2011). Facilitating knowledge transfer and the achievement of competitive advantage with corporate universities: An exploratory model based on media richness and type of knowledge to be transferred. In M. Jennex (Ed.), *Global aspects and cultural perspectives on knowledge management: Emerging dimensions* (pp. 329–345). Hershey, PA: IGI Global. doi:10.4018/978-1-60960-555-1.ch020

Colomb, R. M. (2013). Representation of action is a primary requirement in ontologies for interoperating information systems. In M. Nazir Ahmad, R. Colomb, & M. Abdullah (Eds.), *Ontology-based applications for enterprise systems and knowledge management* (pp. 68–76). Hershey, PA: IGI Global. doi:10.4018/978-1-4666-1993-7.ch004

Colucci, S., Di Noia, T., Di Sciascio, E., Donini, F. M., & Mongiello, M. (2011). Description logic-based resource retrieval. In D. Schwartz & D. Te'eni (Eds.), *Encyclopedia of knowledge management* (2nd ed.; pp. 185–197). Hershey, PA: IGI Global. doi:10.4018/978-1-59904-931-1.ch018

Connell, N. A. (2011). Organisational storytelling. In D. Schwartz & D. Te'eni (Eds.), *Encyclopedia of knowledge management* (2nd ed.; pp. 1261–1269). Hershey, PA: IGI Global. doi:10.4018/978-1-59904-931-1.ch120

Cooper, L. P., & Rober, M. B. (2012). Moving wikis behind the firewall: Intrapedias and work-wikis. In I. Boughzala & A. Dudezert (Eds.), *Knowledge Management 2.0: Organizational Models and Enterprise Strategies* (pp. 44–63). Hershey, PA: IGI Global. doi:10.4018/978-1-61350-195-5.ch003

Corallo, A., De Maggio, M., & Margherita, A. (2011). Knowledge democracy as the new mantra in product innovation: A framework of processes and competencies. In A. Eardley & L. Uden (Eds.), *Innovative knowledge management: Concepts for organizational creativity and collaborative design* (pp. 141–156). Hershey, PA: IGI Global. doi:10.4018/978-1-60566-701-0.ch008

Costa, G. (2011). Knowledge worker fair compensation: Ethical issues and social dilemmas. In G. Morais da Costa (Ed.), *Ethical issues and social dilemmas in knowledge management: Organizational innovation* (pp. 215–231). Hershey, PA: IGI Global. doi:10.4018/978-1-61520-873-9.ch013

Costello, R. (2014). Evaluating e-learning from an end user perspective. In M. Pańkowska (Ed.), *Frameworks of IT prosumption for business development* (pp. 259–283). Hershey, PA: IGI Global. doi:10.4018/978-1-4666-4313-0.ch017

Crasso, M., Zunino, A., & Campo, M. (2013). A survey of approaches to web service discovery in service-oriented architectures. In K. Siau (Ed.), *Innovations in database design, web applications, and information systems management* (pp. 107–138). Hershey, PA: IGI Global. doi:10.4018/978-1-4666-2044-5.ch005

Croasdell, D., & Wang, Y. K. (2011). Virtue-nets. In D. Schwartz & D. Te'eni (Eds.), *Encyclopedia of knowledge management* (2nd ed.; pp. 1545–1555). Hershey, PA: IGI Global. doi:10.4018/978-1-59904-931-1.ch147

Cucchiara, S., Ligorio, M. B., & Fujita, N. (2014). Understanding online discourse strategies for knowledge building through social network analysis. In H. Lim & F. Sudweeks (Eds.), *Innovative methods and technologies for electronic discourse analysis* (pp. 42–62). Hershey, PA: IGI Global. doi:10.4018/978-1-4666-4426-7.ch003

Cudanov, M., & Kirchner, K. (2011). Knowledge management in high-growth companies: A case study in Serbia. In M. Al-Shammari (Ed.), *Knowledge management in emerging economies: Social, organizational and cultural implementation* (pp. 227–248). Hershey, PA: IGI Global. doi:10.4018/978-1-61692-886-5.ch014

Cuel, R., Bouquet, P., & Bonifacio, M. (2011). Distributed knowledge management. In D. Schwartz & D. Te'eni (Eds.), *Encyclopedia of knowledge management* (2nd ed.; pp. 198–208). Hershey, PA: IGI Global. doi:10.4018/978-1-59904-931-1.ch019

Daidj, N. (2012). The evolution of KM practices: The case of the Renault-Nissan international strategic alliance. In I. Boughzala & A. Dudezert (Eds.), *Knowledge management 2.0: Organizational models and enterprise strategies* (pp. 190–213). Hershey, PA: IGI Global. doi:10.4018/978-1-61350-195-5.ch010

Daniel, B. K., Zapata-Rivera, J., & McCalla, G. I. (2010). A Bayesian belief network methodology for modeling social systems in virtual communities: Opportunities for database technologies. In K. Anbumani & R. Nedunchezhian (Eds.), *Soft computing applications for database technologies: Techniques and issues* (pp. 125–152). Hershey, PA: IGI Global. doi:10.4018/978-1-60566-814-7.ch008

Darchen, S., & Tremblay, D. (2010). Attracting and retaining knowledge workers: The impact of quality of place in the case of Montreal. In K. Metaxiotis, F. Carrillo, & T. Yigitcanlar (Eds.), *Knowledge-based development for cities and societies: Integrated multi-level approaches* (pp. 42–58). Hershey, PA: IGI Global. doi:10.4018/978-1-61520-721-3.ch003

Davenport, D. L., & Hosapple, C. W. (2011). Knowledge organizations. In D. Schwartz & D. Te'eni (Eds.), *Encyclopedia of knowledge management* (2nd ed.; pp. 822–832). Hershey, PA: IGI Global. doi:10.4018/978-1-59904-931-1.ch079

Davenport, D. L., & Hosapple, C. W. (2011). Social capital knowledge. In D. Schwartz & D. Te'eni (Eds.), *Encyclopedia of knowledge management* (2nd ed.; pp. 1448–1459). Hershey, PA: IGI Global. doi:10.4018/978-1-59904-931-1.ch138

De Maggio, M., Del Vecchio, P., Elia, G., & Grippa, F. (2011). An ICT-based network of competence centres for developing intellectual capital in the Mediterranean area. In A. Al Ajeeli & Y. Al-Bastaki (Eds.), *Handbook of research on e-services in the public sector: E-government strategies and advancements* (pp. 164–181). Hershey, PA: IGI Global. doi:10.4018/978-1-61520-789-3.ch014

Delbaere, M., Di Zhang, D., Bruning, E. R., & Sivaramakrishnan, S. (2014). Knowledge management and the roles it plays in achieving superior performance. In M. Chilton & J. Bloodgood (Eds.), *Knowledge management and competitive advantage: Issues and potential solutions* (pp. 90–108). Hershey, PA: IGI Global. doi:10.4018/978-1-4666-4679-7.ch006

Delgado-Verde, M., & Cruz-González, J. (2010). An intellectual capital-based view of technological innovation. In P. López Sáez, G. Castro, J. Navas López, & M. Delgado Verde (Eds.), *Intellectual capital and technological innovation: Knowledge-based theory and practice* (pp. 166–193). Hershey, PA: IGI Global. doi:10.4018/978-1-61520-875-3.ch008

Deltour, F., Plé, L., & Roussel, C. S. (2012). Knowledge sharing in the age of web 2.0: A social capital perspective. In I. Boughzala & A. Dudezert (Eds.), *Knowledge management 2.0: Organizational models and enterprise strategies* (pp. 122–141). Hershey, PA: IGI Global. doi:10.4018/978-1-61350-195-5.ch007

Derballa, V., & Pousttchi, K. (2011). Mobile technology for knowledge management. In D. Schwartz & D. Te'eni (Eds.), *Encyclopedia of knowledge management* (2nd ed.; pp. 1158–1166). Hershey, PA: IGI Global. doi:10.4018/978-1-59904-931-1.ch111

Dieng-Kuntz, R. (2011). Corporate semantic webs. In D. Schwartz & D. Te'eni (Eds.), *Encyclopedia of knowledge management* (2nd ed.; pp. 131–149). Hershey, PA: IGI Global. doi:10.4018/978-1-59904-931-1.ch014

Diosteanu, A., Stellato, A., & Turbati, A. (2012). SODA: A service oriented data acquisition framework. In M. Pazienza & A. Stellato (Eds.), *Semi-automatic ontology development: Processes and resources* (pp. 48–77). Hershey, PA: IGI Global. doi:10.4018/978-1-4666-0188-8.ch003

Donate-Manzanares, M. J., Guadamillas-Gómez, F., & Sánchez de Pablo, J. D. (2010). Strategic alliances and knowledge management strategies: A case study. In M. Russ (Ed.), *Knowledge management strategies for business development* (pp. 240–260). Hershey, PA: IGI Global. doi:10.4018/978-1-60566-348-7.ch011

Donnet, T., Keast, R., & Pickernell, D. (2010). Up the junction? Exploiting knowledge-based development through supply chain and SME cluster interactions. In K. Metaxiotis, F. Carrillo, & T. Yigitcanlar (Eds.), *Knowledge-based development for cities and societies: Integrated multi-level approaches* (pp. 179–195). Hershey, PA: IGI Global. doi:10.4018/978-1-61520-721-3.ch011

Douglas, I. (2011). Organizational needs analysis and knowledge management. In D. Schwartz & D. Te'eni (Eds.), *Encyclopedia of knowledge management* (2nd ed.; pp. 1290–1297). Hershey, PA: IGI Global. doi:10.4018/978-1-59904-931-1.ch123

Edvardsson, I. R., & Oskarsson, G. K. (2013). Outsourcing in knowledge-based service firms. In J. Wang (Ed.), *Intelligence methods and systems advancements for knowledge-based business* (pp. 97–113). Hershey, PA: IGI Global. doi:10.4018/978-1-4666-1873-2.ch006

Elenurm, T. (2013). Knowledge management and innovative learning. In S. Buckley & M. Jakovljevic (Eds.), *Knowledge management innovations for interdisciplinary education: Organizational applications* (pp. 108–131). Hershey, PA: IGI Global. doi:10.4018/978-1-4666-1969-2.ch006

Eppler, M. J., & Burkhard, R. A. (2011). Knowledge visualization. In D. Schwartz & D. Te'eni (Eds.), *Encyclopedia of knowledge management* (2nd ed.; pp. 987–999). Hershey, PA: IGI Global. doi:10.4018/978-1-59904-931-1.ch094

Ergazakis, K., Metaxiotis, K., & Ergazakis, E. (2011). Exploring paths towards knowledge cities developments: A research agenda. In D. Schwartz & D. Te'eni (Eds.), *Encyclopedia of knowledge management* (2nd ed.; pp. 288–297). Hershey, PA: IGI Global. doi:10.4018/978-1-59904-931-1.ch028

Eri, Z. D., Abdullah, R., Jabar, M. A., Murad, M. A., & Talib, A. M. (2013). Ontology-based virtual communities model for the knowledge management system environment: Ontology design. In M. Nazir Ahmad, R. Colomb, & M. Abdullah (Eds.), *Ontology-based applications for enterprise systems and knowledge management* (pp. 343–360). Hershey, PA: IGI Global. doi:10.4018/978-1-4666-1993-7.ch019

Erickson, G. S. (2014). Government as a partner in knowledge management: Lessons from the US freedom of information act. In Y. Al-Bastaki & A. Shajera (Eds.), *Building a competitive public sector with knowledge management strategy* (pp. 90–103). Hershey, PA: IGI Global. doi:10.4018/978-1-4666-4434-2.ch003

Erickson, G. S., & Rothberg, H. N. (2011). Assessing knowledge management needs: A strategic approach to developing knowledge. *International Journal of Knowledge Management, 7*(3), 1–10. doi:10.4018/jkm.2011070101

Erickson, G. S., & Rothberg, H. N. (2011). Protecting knowledge assets. In D. Schwartz & D. Te'eni (Eds.), *Encyclopedia of knowledge management* (2nd ed.; pp. 1336–1342). Hershey, PA: IGI Global. doi:10.4018/978-1-59904-931-1.ch128

Erickson, G. S., & Rothberg, H. N. (2013). Assessing knowledge management needs: A strategic approach to developing knowledge. In M. Jennex (Ed.), *Dynamic models for knowledge-driven organizations* (pp. 180–189). Hershey, PA: IGI Global. doi:10.4018/978-1-4666-2485-6.ch010

Evermann, J., & Wand, Y. (2011). Ontology based object-oriented domain modeling: Representing behavior. In K. Siau (Ed.), *Theoretical and practical advances in information systems development: Emerging trends and approaches* (pp. 37–60). Hershey, PA: IGI Global. doi:10.4018/978-1-60960-521-6.ch003

Fadel, K. J., Durcikova, A., & Cha, H. S. (2011). An experiment of information elaboration in mediated knowledge transfer. In M. Jennex (Ed.), *Global aspects and cultural perspectives on knowledge management: Emerging dimensions* (pp. 311–328). Hershey, PA: IGI Global. doi:10.4018/978-1-60960-555-1.ch019

Fazel-Zarandi, M., Fox, M. S., & Yu, E. (2013). Ontologies in expertise finding systems: Modeling, analysis, and design. In M. Nazir Ahmad, R. Colomb, & M. Abdullah (Eds.), *Ontology-based applications for enterprise systems and knowledge management* (pp. 158–177). Hershey, PA: IGI Global. doi:10.4018/978-1-4666-1993-7.ch009

Ferri, F., & Grifoni, P. (2011). Sketching in knowledge creation and management. In D. Schwartz & D. Te'eni (Eds.), *Encyclopedia of knowledge management* (2nd ed.; pp. 1438–1447). Hershey, PA: IGI Global. doi:10.4018/978-1-59904-931-1.ch137

Filho, C. G., Baroni de Carvalho, R., & Jamil, G. L. (2011). Market knowledge management, innovation and product performance: Survey in medium and large Brazilian industrial firms. In M. Jennex & S. Smolnik (Eds.), *Strategies for knowledge management success: Exploring organizational efficacy* (pp. 32–50). Hershey, PA: IGI Global. doi:10.4018/978-1-60566-709-6.ch003

Fink, D., & Disterer, G. (2011). Knowledge management in professional service firms. In D. Schwartz & D. Te'eni (Eds.), *Encyclopedia of knowledge management* (2nd ed.; pp. 650–659). Hershey, PA: IGI Global. doi:10.4018/978-1-59904-931-1.ch063

Fink, K. (2011). Process model for knowledge potential measurement in SMEs. In M. Jennex & S. Smolnik (Eds.), *Strategies for knowledge management success: Exploring organizational efficacy* (pp. 91–105). Hershey, PA: IGI Global. doi:10.4018/978-1-60566-709-6.ch006

Fink, K., & Ploder, C. (2011). Knowledge management toolkit for SMEs. In M. Jennex (Ed.), *Global aspects and cultural perspectives on knowledge management: Emerging dimensions* (pp. 49–63). Hershey, PA: IGI Global. doi:10.4018/978-1-60960-555-1.ch004

Flynn, R., & Marshall, V. (2014). The four levers for change in knowledge management implementation. In Y. Al-Bastaki & A. Shajera (Eds.), *Building a competitive public sector with knowledge management strategy* (pp. 227–245). Hershey, PA: IGI Global. doi:10.4018/978-1-4666-4434-2.ch010

Fortier, J., & Kassel, G. (2011). Organizational semantic webs. In D. Schwartz & D. Te'eni (Eds.), *Encyclopedia of knowledge management* (2nd ed.; pp. 1298–1307). Hershey, PA: IGI Global. doi:10.4018/978-1-59904-931-1.ch124

Framinan, J. M., & Molina, J. M. (2010). An overview of enterprise resource planning for intelligent enterprises. In *Business information systems: Concepts, methodologies, tools and applications* (pp. 60–68). Hershey, PA: IGI Global. doi:10.4018/978-1-61520-969-9.ch005

Franco, M., Di Virgilio, F., & Di Pietro, L. (2014). Management of group knowledge and the role of E-WOM for business organizations. In M. Chilton & J. Bloodgood (Eds.), *Knowledge management and competitive advantage: Issues and potential solutions* (pp. 71–89). Hershey, PA: IGI Global. doi:10.4018/978-1-4666-4679-7.ch005

Franke, U. J. (2000). The knowledge-based view (KBV) of the virtual web, the virtual corporation and the net-broker. In Y. Malhotra (Ed.), *Knowledge management and virtual organizations* (pp. 20–42). Hershey, PA: IGI Global. doi:10.4018/978-1-930708-65-5.ch002

Freivalds, D., & Lush, B. (2012). Thinking inside the grid: Selecting a discovery system through the RFP process. In M. Popp & D. Dallis (Eds.), *Planning and implementing resource discovery tools in academic libraries* (pp. 104–121). Hershey, PA: IGI Global. doi:10.4018/978-1-4666-1821-3.ch007

Frieß, M. R., Groh, G., Reinhardt, M., Forster, F., & Schlichter, J. (2012). Context-aware creativity support for corporate open innovation. *International Journal of Knowledge-Based Organizations, 2*(1), 38–55. doi:10.4018/ijkbo.2012010103

Fuller, C. M., & Wilson, R. L. (2011). Extracting knowledge from neural networks. In D. Schwartz & D. Te'eni (Eds.), *Encyclopedia of knowledge management* (2nd ed.; pp. 320–330). Hershey, PA: IGI Global. doi:10.4018/978-1-59904-931-1.ch031

Furquim, T. D., & do Amaral, S. A. (2011). Knowledge management practices in brazilian software organizations: The case of SERPRO. In M. Al-Shammari (Ed.), *Knowledge management in emerging economies: Social, organizational and cultural implementation* (pp. 213–226). Hershey, PA: IGI Global. doi:10.4018/978-1-61692-886-5.ch013

Gaál, Z., Szabó, L., Obermayer-Kovács, N., Kovács, Z., & Csepregi, A. (2011). Knowledge management profile: An innovative approach to map knowledge management practice. In A. Eardley & L. Uden (Eds.), *Innovative knowledge management: Concepts for organizational creativity and collaborative design* (pp. 253–263). Hershey, PA: IGI Global. doi:10.4018/978-1-60566-701-0.ch016

Ganguly, A., Mostashari, A., & Mansouri, M. (2013). Measuring knowledge management/knowledge sharing (KM/KS) efficiency and effectiveness in enterprise networks. In M. Jennex (Ed.), *Dynamic models for knowledge-driven organizations* (pp. 318–336). Hershey, PA: IGI Global. doi:10.4018/978-1-4666-2485-6.ch019

Gaumand, C., Chapdaniel, A., & Dudezert, A. (2012). Strategic knowledge management system framework for supply chain at an intra-organizational level. In I. Boughzala & A. Dudezert (Eds.), *Knowledge management 2.0: Organizational models and enterprise strategies* (pp. 142–163). Hershey, PA: IGI Global. doi:10.4018/978-1-61350-195-5.ch008

Ghazali, R., & Zakaria, N. H. (2013). Knowledge management processes in enterprise systems: A systematic literature review. In M. Nazir Ahmad, R. Colomb, & M. Abdullah (Eds.), *Ontology-based applications for enterprise systems and knowledge management* (pp. 1–24). Hershey, PA: IGI Global. doi:10.4018/978-1-4666-1993-7.ch001

Ghosh, B. (2011). Cross-cultural knowledge management practices to support offshore outsourcing. In M. Al-Shammari (Ed.), *Knowledge management in emerging economies: Social, organizational and cultural implementation* (pp. 249–260). Hershey, PA: IGI Global. doi:10.4018/978-1-61692-886-5.ch015

Gohil, U., Carrillo, P., Ruikar, K., & Anumba, C. (2013). Development of a business process model for a project-based service organisation. *International Journal of Knowledge-Based Organizations, 3*(1), 37–56. doi:10.4018/ijkbo.2013010103

Goldsmith, R. E., & Pillai, K. G. (2011). Knowledge calibration and knowledge management. In D. Schwartz & D. Te'eni (Eds.), *Encyclopedia of knowledge management* (2nd ed.; pp. 497–505). Hershey, PA: IGI Global. doi:10.4018/978-1-59904-931-1.ch048

Gomes de Andrade, F., & Baptista, C. D. (2013). An ontology-based approach to support information discovery in spatial data infrastructures. In C. Rückemann (Ed.), *Integrated information and computing systems for natural, spatial, and social sciences* (pp. 369–387). Hershey, PA: IGI Global. doi:10.4018/978-1-4666-2190-9.ch018

Gonçalo, C. R., & Jacques, E. J. (2010). Best practices of knowledge strategy in hospitals: A contextual perspective based on the implementation of medical protocols. In D. Harorimana (Ed.), *Cultural implications of knowledge sharing, management and transfer: Identifying competitive advantage* (pp. 180–202). Hershey, PA: IGI Global. doi:10.4018/978-1-60566-790-4.ch009

Górniak-Kocikowska, K. (2011). Knowledge management and democracy: A critical review of some moral issues and social dilemmas. In G. Morais da Costa (Ed.), *Ethical issues and social dilemmas in knowledge management: Organizational innovation* (pp. 28–44). Hershey, PA: IGI Global. doi:10.4018/978-1-61520-873-9.ch003

Gottschalk, P. (2014). Police knowledge management strategy. In M. Chilton & J. Bloodgood (Eds.), *Knowledge management and competitive advantage: Issues and potential solutions* (pp. 202–220). Hershey, PA: IGI Global. doi:10.4018/978-1-4666-4679-7.ch011

Goudos, S. K., Peristeras, V., & Tarabanis, K. (2010). Application of semantic web technology in e-business: Case studies in public domain data knowledge representation. In *Business information systems: Concepts, methodologies, tools and applications* (pp. 1223–1233). Hershey, PA: IGI Global. doi:10.4018/978-1-61520-969-9.ch075

Govindarajan, M., & Chandrasekaran, R. (2012). A hybrid multilayer perceptron neural network for direct marketing. *International Journal of Knowledge-Based Organizations*, *2*(3), 63–73. doi:10.4018/ijkbo.2012070104

Grant, J., & Minker, J. (2011). Logic and knowledge bases. In D. Schwartz & D. Te'eni (Eds.), *Encyclopedia of knowledge management* (2nd ed.; pp. 1022–1033). Hershey, PA: IGI Global. doi:10.4018/978-1-59904-931-1.ch097

Green, A. (2011). Engineering business reasoning, analytics and intelligence network (E-BRAIN): A new approach to intangible asset valuation based on Einstein's perspective. In B. Vallejo-Alonso, A. Rodriguez-Castellanos, & G. Arregui-Ayastuy (Eds.), *Identifying, measuring, and valuing knowledge-based intangible assets: New perspectives* (pp. 232–253). Hershey, PA: IGI Global. doi:10.4018/978-1-60960-054-9.ch011

Greenaway, K. E., & Vuong, D. C. (2010). Taking charities seriously: A call for focused knowledge management research. *International Journal of Knowledge Management*, *6*(4), 87–97. doi:10.4018/jkm.2010100105

Greenaway, K. E., & Vuong, D. C. (2012). Taking charities seriously: A call for focused knowledge management research. In M. Jennex (Ed.), *Conceptual models and outcomes of advancing knowledge management: New technologies* (pp. 333–344). Hershey, PA: IGI Global. doi:10.4018/978-1-4666-0035-5.ch017

Gunjal, B., Gaitanou, P., & Yasin, S. (2012). Social networks and knowledge management: An explorative study in library systems. In I. Boughzala & A. Dudezert (Eds.), *Knowledge management 2.0: Organizational models and enterprise strategies* (pp. 64–83). Hershey, PA: IGI Global. doi:10.4018/978-1-61350-195-5.ch004

Habhab-Rave, S. (2010). Knowledge management in SMEs: A mixture of innovation, marketing and ICT: Analysis of two case studies. In E. O'Brien, S. Clifford, & M. Southern (Eds.), *Knowledge management for process, organizational and marketing innovation: Tools and methods* (pp. 183–194). Hershey, PA: IGI Global. doi:10.4018/978-1-61520-829-6.ch011

Habicht, H., Möslein, K. M., & Reichwald, R. (2012). Open innovation maturity. *International Journal of Knowledge-Based Organizations, 2*(1), 92–111. doi:10.4018/ijkbo.2012010106

Hamburg, I., & Hall, T. (2010). Readiness for knowledge management, methods and environments for innovation. In E. O'Brien, S. Clifford, & M. Southern (Eds.), *Knowledge management for process, organizational and marketing innovation: Tools and methods* (pp. 1–15). Hershey, PA: IGI Global. doi:10.4018/978-1-61520-829-6.ch001

Hamza, S. E. (2011). Capturing tacit knowledge from transient workers: Improving the organizational competitiveness. In M. Jennex (Ed.), *Global aspects and cultural perspectives on knowledge management: Emerging dimensions* (pp. 172–188). Hershey, PA: IGI Global. doi:10.4018/978-1-60960-555-1.ch012

Harorimana, D. (2010). Knowledge, culture, and cultural impact on knowledge management: Some lessons for researchers and practitioners. In D. Harorimana (Ed.), *Cultural implications of knowledge sharing, management and transfer: Identifying competitive advantage* (pp. 48–59). Hershey, PA: IGI Global. doi:10.4018/978-1-60566-790-4.ch003

Hasan, H. (2011). Formal and emergent standards in KM. In D. Schwartz & D. Te'eni (Eds.), *Encyclopedia of knowledge management* (2nd ed.; pp. 331–342). Hershey, PA: IGI Global. doi:10.4018/978-1-59904-931-1.ch032

He, G., Xue, G., Yu, K., & Yao, S. (2013). Business process modeling: Analysis and evaluation. In Z. Lu (Ed.), *Design, performance, and analysis of innovative information retrieval* (pp. 382–393). Hershey, PA: IGI Global. doi:10.4018/978-1-4666-1975-3.ch027

Heiman, B. A., & Hurmelinna-Laukkanen, P. (2010). Problem finding and solving: A knowledge-based view of managing innovation. In P. López Sáez, G. Castro, J. Navas López, & M. Delgado Verde (Eds.), *Intellectual capital and technological innovation: Knowledge-based theory and practice* (pp. 105–130). Hershey, PA: IGI Global. doi:10.4018/978-1-61520-875-3.ch005

Hendriks, P. H. (2011). Organizational structure. In D. Schwartz & D. Te'eni (Eds.), *Encyclopedia of knowledge management* (2nd ed.; pp. 1308–1318). Hershey, PA: IGI Global. doi:10.4018/978-1-59904-931-1.ch125

Hercheui, M. D. (2012). KMS for fostering behavior change: A case study on Microsoft Hohm. In I. Boughzala & A. Dudezert (Eds.), *Knowledge management 2.0: Organizational models and enterprise strategies* (pp. 214–232). Hershey, PA: IGI Global. doi:10.4018/978-1-61350-195-5.ch011

Hipkin, I. (2011). Perceptions of factors influencing knowledge-based technology management in conflict areas. In M. Al-Shammari (Ed.), *Knowledge management in emerging economies: Social, organizational and cultural implementation* (pp. 294–307). Hershey, PA: IGI Global. doi:10.4018/978-1-61692-886-5.ch018

Hofer, F. (2011). Knowledge transfer between academia and industry. In D. Schwartz & D. Te'eni (Eds.), *Encyclopedia of knowledge management* (2nd ed.; pp. 977–986). Hershey, PA: IGI Global. doi:10.4018/978-1-59904-931-1.ch093

Holjevac, I. A., Crnjar, K., & Hrgovic, A. V. (2013). Knowledge management and quality in Croatian tourism. In S. Buckley & M. Jakovljevic (Eds.), *Knowledge management innovations for interdisciplinary education: Organizational applications* (pp. 178–192). Hershey, PA: IGI Global. doi:10.4018/978-1-4666-1969-2.ch009

Holsapple, C. W., & Joshi, K. D. (2011). Knowledge management ontology. In D. Schwartz & D. Te'eni (Eds.), *Encyclopedia of knowledge management* (2nd ed.; pp. 704–711). Hershey, PA: IGI Global. doi:10.4018/978-1-59904-931-1.ch068

Holsapple, C. W., & Oh, J. (2014). Reactive and proactive dynamic capabilities: Using the knowledge chain theory of competitiveness. In M. Chilton & J. Bloodgood (Eds.), *Knowledge management and competitive advantage: Issues and potential solutions* (pp. 1–19). Hershey, PA: IGI Global. doi:10.4018/978-1-4666-4679-7.ch001

Huang, A., Xiao, J., & Wang, S. (2013). A combined forecast method integrating contextual knowledge. In G. Yang (Ed.), *Multidisciplinary studies in knowledge and systems science* (pp. 274–290). Hershey, PA: IGI Global. doi:10.4018/978-1-4666-3998-0.ch019

Huff, C. (2011). What does knowledge have to do with ethics? In G. Morais da Costa (Ed.), *Ethical issues and social dilemmas in knowledge management: Organizational innovation* (pp. 17–27). Hershey, PA: IGI Global. doi:10.4018/978-1-61520-873-9.ch002

Hürster, W., Wilbois, T., & Chaves, F. (2010). An integrated systems approach for early warning and risk management systems. *International Journal of Information Technologies and Systems Approach, 3*(2), 46–56. doi:10.4018/jitsa.2010070104

Iyer, S. R., Sharda, R., Biros, D., Lucca, J., & Shimp, U. (2011). Organization of lessons learned knowledge: A taxonomy and implementation. In M. Jennex (Ed.), *Global aspects and cultural perspectives on knowledge management: Emerging dimensions* (pp. 190–209). Hershey, PA: IGI Global. doi:10.4018/978-1-60960-555-1.ch013

Jacobson, C. M. (2011). Knowledge sharing between individuals. In D. Schwartz & D. Te'eni (Eds.), *Encyclopedia of knowledge management* (2nd ed.; pp. 924–934). Hershey, PA: IGI Global. doi:10.4018/978-1-59904-931-1.ch088

Jakovljevic, M. (2013). A conceptual model of creativity, invention, and innovation (MCII) for entrepreneurial engineers. In S. Buckley & M. Jakovljevic (Eds.), *Knowledge management innovations for interdisciplinary education: Organizational applications* (pp. 66–87). Hershey, PA: IGI Global. doi:10.4018/978-1-4666-1969-2.ch004

Jasimuddin, S. M., Connell, N., & Klein, J. H. (2011). Understanding organizational memory. In D. Schwartz & D. Te'eni (Eds.), *Encyclopedia of knowledge management* (2nd ed.; pp. 1536–1544). Hershey, PA: IGI Global. doi:10.4018/978-1-59904-931-1.ch146

Jennex, M. E. (2010). Do organizational memory and information technology interact to affect organizational information needs and provision? In *Ubiquitous developments in knowledge management: Integrations and trends* (pp. 1–20). Hershey, PA: IGI Global. doi:10.4018/978-1-60566-954-0

Jennex, M. E. (2010). Knowledge sharing model of 24-hour knowledge factory. In *Ubiquitous developments in knowledge management: Integrations and trends* (pp. 141–154). Hershey, PA: IGI Global. doi:10.4018/978-1-60566-954-0

Jennex, M. E. (2010). Operationalizing knowledge sharing for informers. In *Ubiquitous developments in knowledge management: Integrations and trends* (pp. 319–340). Hershey, PA: IGI Global. doi:10.4018/978-1-60566-954-0

Jennex, M. E. (2010). Qualitative pre-processing for semantic search of unstructured knowledge. In *Ubiquitous developments in knowledge management: Integrations and trends* (pp. 252–263). Hershey, PA: IGI Global. doi:10.4018/978-1-60566-954-0

Jennex, M. E. (2010). A specialized evaluation and comparison of sample data mining software. In *Ubiquitous developments in knowledge management: Integrations and trends* (pp. 300–318). Hershey, PA: IGI Global. doi:10.4018/978-1-60566-954-0

Jennex, M. E. (2010). Using soft systems methodology to reveal socio-technical barriers to knowledge sharing and management: A case study from the UK national health service. In *Ubiquitous developments in knowledge management: Integrations and trends* (pp. 215–235). Hershey, PA: IGI Global. doi:10.4018/978-1-60566-954-0

Jennex, M. E. (2011). Knowledge management success models. In D. Schwartz & D. Te'eni (Eds.), *Encyclopedia of knowledge management* (2nd ed.; pp. 763–771). Hershey, PA: IGI Global. doi:10.4018/978-1-59904-931-1.ch074

Jennex, M. E., & Olfman, L. (2011). A model of knowledge management success. In M. Jennex & S. Smolnik (Eds.), *Strategies for knowledge management success: Exploring organizational efficacy* (pp. 14–31). Hershey, PA: IGI Global. doi:10.4018/978-1-60566-709-6.ch002

Jennex, M. E., Smolnik, S., & Croasdell, D. (2011). Towards a consensus knowledge management success definition. In M. Jennex & S. Smolnik (Eds.), *Strategies for knowledge management success: Exploring organizational efficacy* (pp. 1–13). Hershey, PA: IGI Global. doi:10.4018/978-1-60566-709-6.ch001

Jewels, T. (2013). Teaching enterprise information systems in the United Arab Emirates. In F. Albadri (Ed.), *Information systems applications in the Arab education sector* (pp. 322–337). Hershey, PA: IGI Global. doi:10.4018/978-1-4666-1984-5.ch022

Jolly, R., & Wakeland, W. (2011). Using agent based simulation and game theory analysis to study knowledge flow in organizations: The KMscape. In M. Jennex (Ed.), *Global aspects and cultural perspectives on knowledge management: Emerging dimensions* (pp. 19–29). Hershey, PA: IGI Global. doi:10.4018/978-1-60960-555-1.ch002

Joshi, S. (2014). Web 2.0 and its implications on globally competitive business model. In M. Pańkowska (Ed.), *Frameworks of IT prosumption for business development* (pp. 86–101). Hershey, PA: IGI Global. doi:10.4018/978-1-4666-4313-0.ch007

Judge, R. (2011). A simulation system for evaluating knowledge management system (KMS) implementation strategies in small to mid-size enterprises (SME). In M. Jennex (Ed.), *Global aspects and cultural perspectives on knowledge management: Emerging dimensions* (pp. 92–112). Hershey, PA: IGI Global. doi:10.4018/978-1-60960-555-1.ch007

Kalid, K. S. (2011). Transfer knowledge using stories: A Malaysian university case study. In M. Al-Shammari (Ed.), *Knowledge management in emerging economies: Social, organizational and cultural implementation* (pp. 186–198). Hershey, PA: IGI Global. doi:10.4018/978-1-61692-886-5.ch011

Kamau, C. (2010). Strategising impression management in corporations: Cultural knowledge as capital. In D. Harorimana (Ed.), *Cultural implications of knowledge sharing, management and transfer: Identifying competitive advantage* (pp. 60–83). Hershey, PA: IGI Global. doi:10.4018/978-1-60566-790-4.ch004

Kamthan, P., & Fancott, T. (2011). A knowledge management model for patterns. In D. Schwartz & D. Te'eni (Eds.), *Encyclopedia of knowledge management* (2nd ed.; pp. 694–703). Hershey, PA: IGI Global. doi:10.4018/978-1-59904-931-1.ch067

Kamthan, P., & Pai, H. (2011). Knowledge representation in pattern management. In D. Schwartz & D. Te'eni (Eds.), *Encyclopedia of knowledge management* (2nd ed.; pp. 893–904). Hershey, PA: IGI Global. doi:10.4018/978-1-59904-931-1.ch085

Kane, G. C., Schwaig, K. S., & Storey, V. C. (2011). Information privacy: Understanding how firms behave online. In K. Siau (Ed.), *Theoretical and practical advances in information systems development: Emerging trends and approaches* (pp. 81–100). Hershey, PA: IGI Global. doi:10.4018/978-1-60960-521-6.ch005

Kankanhalli, A., Tan, B. C., & Wei, K. (2011). Knowledge producers and consumers. In D. Schwartz & D. Te'eni (Eds.), *Encyclopedia of knowledge management* (2nd ed.; pp. 867–877). Hershey, PA: IGI Global. doi:10.4018/978-1-59904-931-1.ch083

Karagiannis, D., Woitsch, R., & Hrgovcic, V. (2010). Industrialisation of the knowledge work: The knowledge conveyer belt approach. In E. O'Brien, S. Clifford, & M. Southern (Eds.), *Knowledge management for process, organizational and marketing innovation: Tools and methods* (pp. 79–94). Hershey, PA: IGI Global. doi:10.4018/978-1-61520-829-6.ch005

Karlsson, F., & Ågerfalk, P. J. (2011). Towards structured flexibility in information systems development: Devising a method for method configuration. In K. Siau (Ed.), *Theoretical and practical advances in information systems development: Emerging trends and approaches* (pp. 214–238). Hershey, PA: IGI Global. doi:10.4018/978-1-60960-521-6.ch010

Karna, A., Singh, R., & Verma, S. (2010). Knowledge management for an effective sales and marketing function. In M. Russ (Ed.), *Knowledge management strategies for business development* (pp. 324–337). Hershey, PA: IGI Global. doi:10.4018/978-1-60566-348-7.ch015

Kassim, A. M., & Cheah, Y. (2013). SEMblog: An ontology-based semantic blogging tool for knowledge identification, organization, and reuse. In M. Nazir Ahmad, R. Colomb, & M. Abdullah (Eds.), *Ontology-based applications for enterprise systems and knowledge management* (pp. 210–223). Hershey, PA: IGI Global. doi:10.4018/978-1-4666-1993-7.ch012

Kayakutlu, G. (2010). Knowledge worker profile: A framework to clarify expectations. In K. Metaxiotis, F. Carrillo, & T. Yigitcanlar (Eds.), *Knowledge-based development for cities and societies: Integrated multi-level approaches* (pp. 162–178). Hershey, PA: IGI Global. doi:10.4018/978-1-61520-721-3.ch010

Kettunen, J., & Chaudhuri, M. R. (2011). Knowledge management to promote organizational change in India. In M. Al-Shammari (Ed.), *Knowledge management in emerging economies: Social, organizational and cultural implementation* (pp. 308–324). Hershey, PA: IGI Global. doi:10.4018/978-1-61692-886-5.ch019

Khalil, O. E., & Seleim, A. (2012). Culture and knowledge transfer capacity: A cross-national study. In M. Jennex (Ed.), *Conceptual models and outcomes of advancing knowledge management: New technologies* (pp. 305–332). Hershey, PA: IGI Global. doi:10.4018/978-1-4666-0035-5.ch016

Khasawneh, R., & Alazzam, A. (2014). Towards customer knowledge management (CKM): Where knowledge and customer meet. In M. Chilton & J. Bloodgood (Eds.), *Knowledge management and competitive advantage: Issues and potential solutions* (pp. 109–121). Hershey, PA: IGI Global. doi:10.4018/978-1-4666-4679-7.ch007

Kim, J. (2014). Big data sharing among academics. In W. Hu & N. Kaabouch (Eds.), *Big data management, technologies, and applications* (pp. 177–194). Hershey, PA: IGI Global. doi:10.4018/978-1-4666-4699-5.ch008

Kim, S., Felan, J., & Kang, M. H. (2011). An ontological approach to enterprise knowledge modeling in a shipping company. *International Journal of Knowledge Management*, *7*(4), 70–84. doi:10.4018/jkm.2011100105

Kim, S., Felan, J., & Kang, M. H. (2013). An ontological approach to enterprise knowledge modeling in a shipping company. In M. Jennex (Ed.), *Dynamic models for knowledge-driven organizations* (pp. 351–363). Hershey, PA: IGI Global. doi:10.4018/978-1-4666-2485-6.ch021

King, W. R. (2011). Knowledge transfer. In D. Schwartz & D. Te'eni (Eds.), *Encyclopedia of knowledge management* (2nd ed.; pp. 967–976). Hershey, PA: IGI Global. doi:10.4018/978-1-59904-931-1.ch092

Kivijärvi, H., Piirainen, K., & Tuominen, M. (2010). Sustaining organizational innovativeness: Advancing knowledge sharing during the scenario process. *International Journal of Knowledge Management*, *6*(2), 22–39. doi:10.4018/jkm.2010040102

Kivijärvi, H., Piirainen, K., & Tuominen, M. (2012). Sustaining organizational innovativeness: Advancing knowledge sharing during the scenario process. In M. Jennex (Ed.), *Conceptual models and outcomes of advancing knowledge management: New technologies* (pp. 99–117). Hershey, PA: IGI Global. doi:10.4018/978-1-4666-0035-5.ch006

Knyazhansky, M., & Plotkin, T. (2012). Knowledge bases over algebraic models: Some notes about informational equivalence. *International Journal of Knowledge Management*, *8*(1), 22–39. doi:10.4018/jkm.2012010102

Kong, E. (2014). The role of social intelligence in acquiring external knowledge for human capital development, organisational learning, and innovation. In M. Chilton & J. Bloodgood (Eds.), *Knowledge management and competitive advantage: Issues and potential solutions* (pp. 53–70). Hershey, PA: IGI Global. doi:10.4018/978-1-4666-4679-7.ch004

Kor, A., & Orange, G. (2011). A survey of epistemology and its implications on an organisational information and knowledge management model. In A. Eardley & L. Uden (Eds.), *Innovative knowledge management: Concepts for organizational creativity and collaborative design* (pp. 95–124). Hershey, PA: IGI Global. doi:10.4018/978-1-60566-701-0.ch006

Kostrzewa, A., Laaksoharju, M., & Kavathatzopoulos, I. (2011). Management of moral knowledge and ethical processes in organizations. In G. Morais da Costa (Ed.), *Ethical issues and social dilemmas in knowledge management: Organizational innovation* (pp. 199–214). Hershey, PA: IGI Global. doi:10.4018/978-1-61520-873-9.ch012

Kraaijenbrink, J., & Wijnhoven, F. (2011). External knowledge integration. In D. Schwartz & D. Te'eni (Eds.), *Encyclopedia of knowledge management* (2nd ed.; pp. 308–319). Hershey, PA: IGI Global. doi:10.4018/978-1-59904-931-1.ch030

Kraft, T. A., & Steenkamp, A. L. (2012). A holistic approach for understanding project management. In F. Stowell (Ed.), *Systems approach applications for developments in information technology* (pp. 25–39). Hershey, PA: IGI Global. doi:10.4018/978-1-4666-1562-5.ch003

Kulkarni, U., & Freeze, R. (2011). Measuring knowledge management capabilities. In D. Schwartz & D. Te'eni (Eds.), *Encyclopedia of knowledge management* (2nd ed.; pp. 1090–1100). Hershey, PA: IGI Global. doi:10.4018/978-1-59904-931-1.ch104

Kumar, A. S., Alrabea, A., & Sekhar, P. C. (2013). Temporal association rule mining in large databases. In *Data mining: Concepts, methodologies, tools, and applications* (pp. 586–602). Hershey, PA: IGI Global. doi:10.4018/978-1-4666-2455-9.ch029

Laihonen, H., & Koivuaho, M. (2011). Knowledge flow audit: indentifying, measuring and managing knowledge asset dynamics. In B. Vallejo-Alonso, A. Rodriguez-Castellanos, & G. Arregui-Ayastuy (Eds.), *Identifying, measuring, and valuing knowledge-based intangible assets: New perspectives* (pp. 22–42). Hershey, PA: IGI Global. doi:10.4018/978-1-60960-054-9.ch002

Land, F., Amjad, U., & Nolas, S. (2011). Knowledge management processes. In D. Schwartz & D. Te'eni (Eds.), *Encyclopedia of knowledge management* (2nd ed.; pp. 719–727). Hershey, PA: IGI Global. doi:10.4018/978-1-59904-931-1.ch070

Lavanderos, L. P., & Fiol, E. S. (2011). Production cognitive capital as a measurement of intellectual capital. In B. Vallejo-Alonso, A. Rodriguez-Castellanos, & G. Arregui-Ayastuy (Eds.), *Identifying, measuring, and valuing knowledge-based intangible assets: New perspectives* (pp. 112–132). Hershey, PA: IGI Global. doi:10.4018/978-1-60960-054-9.ch006

Lavoué, É., George, S., & Prévôt, P. (2011). A knowledge management tool for the interconnection of communities of practice. *International Journal of Knowledge Management, 7*(1), 55–76. doi:10.4018/jkm.2011010104

Lee, H., Chan, K., & Tsui, E. (2013). Knowledge mining Wikipedia: An ontological approach. In G. Yang (Ed.), *Multidisciplinary studies in knowledge and systems science* (pp. 52–62). Hershey, PA: IGI Global. doi:10.4018/978-1-4666-3998-0.ch005

Leung, N. K. (2011). A re-distributed knowledge management framework in help desk. In D. Schwartz & D. Te'eni (Eds.), *Encyclopedia of knowledge management* (2nd ed.; pp. 1374–1381). Hershey, PA: IGI Global. doi:10.4018/978-1-59904-931-1.ch131

Li, Y., Guo, H., & Wang, S. (2010). A multiple-bits watermark for relational data. In K. Siau & J. Erickson (Eds.), *Principle advancements in database management technologies: New applications and frameworks* (pp. 1–22). Hershey, PA: IGI Global. doi:10.4018/978-1-60566-904-5.ch001

Lin, C. Y. (2013). Intellectual capital explains a country's resilience to financial crisis: A resource-based view. In P. Ordóñez de Pablos, R. Tennyson, & J. Zhao (Eds.), *Intellectual capital strategy management for knowledge-based organizations* (pp. 52–75). Hershey, PA: IGI Global. doi:10.4018/978-1-4666-3655-2.ch005

Lin, Y., & Dalkir, K. (2012). Factors affecting KM implementation in the Chinese community. In M. Jennex (Ed.), *Conceptual models and outcomes of advancing knowledge management: New technologies* (pp. 1–23). Hershey, PA: IGI Global. doi:10.4018/978-1-4666-0035-5.ch001

Lindsey, K. L. (2011). Barriers to knowledge sharing. In D. Schwartz & D. Te'eni (Eds.), *Encyclopedia of knowledge management* (2nd ed.; pp. 49–61). Hershey, PA: IGI Global. doi:10.4018/978-1-59904-931-1.ch006

Liu, K., Tan, H. B., & Chen, X. (2013). Aiding maintenance of database applications through extracting attribute dependency graph. *Journal of Database Management, 24*(1), 20–35. doi:10.4018/jdm.2013010102

Liu, K., Tan, H. B., & Chen, X. (2013). Automated insertion of exception handling for key and referential constraints. *Journal of Database Management, 24*(1), 1–19. doi:10.4018/jdm.2013010101

Locuratolo, E., & Palomäki, J. (2013). Ontology for database preservation. In M. Nazir Ahmad, R. Colomb, & M. Abdullah (Eds.), *Ontology-based applications for enterprise systems and knowledge management* (pp. 141–157). Hershey, PA: IGI Global. doi:10.4018/978-1-4666-1993-7.ch008

López-Nicolás, C., & Meroño-Cerdán, Á. L. (2010). A model for knowledge management and intellectual capital audits. In M. Russ (Ed.), *Knowledge management strategies for business development* (pp. 115–131). Hershey, PA: IGI Global. doi:10.4018/978-1-60566-348-7.ch006

Luck, D. (2010). The implications of the development and implementation of CRM for knowledge management. In M. Russ (Ed.), *Knowledge management strategies for business development* (pp. 338–352). Hershey, PA: IGI Global. doi:10.4018/978-1-60566-348-7.ch016

Lukovic, I., Ivancevic, V., Celikovic, M., & Aleksic, S. (2014). DSLs in action with model based approaches to information system development. In *Software design and development: Concepts, methodologies, tools, and applications* (pp. 596–626). Hershey, PA: IGI Global. doi:10.4018/978-1-4666-4301-7.ch029

Luna-Reyes, L. F., & Gil-Garcia, J. R. (2012). Government and inter-organizational collaboration as strategies for administrative reform in Mexico. In T. Papadopoulos & P. Kanellis (Eds.), *Public sector reform using information technologies: Transforming policy into practice* (pp. 79–101). Hershey, PA: IGI Global. doi:10.4018/978-1-60960-839-2.ch005

Lungu, C. I., Caraiani, C., & Dascalu, C. (2013). Sustainable intellectual capital: The inference of corporate social responsibility within intellectual capital. In P. Ordóñez de Pablos, R. Tennyson, & J. Zhao (Eds.), *Intellectual capital strategy management for knowledge-based organizations* (pp. 156–173). Hershey, PA: IGI Global. doi:10.4018/978-1-4666-3655-2.ch009

Ma, Z. M. (2011). Engineering design knowledge management. In D. Schwartz & D. Te'eni (Eds.), *Encyclopedia of knowledge management* (2nd ed.; pp. 263–269). Hershey, PA: IGI Global. doi:10.4018/978-1-59904-931-1.ch025

Maier, R., & Hadrich, T. (2011). Knowledge management systems. In D. Schwartz & D. Te'eni (Eds.), *Encyclopedia of knowledge management* (2nd ed.; pp. 779–790). Hershey, PA: IGI Global. doi:10.4018/978-1-59904-931-1.ch076

Maria, E. D., & Micelli, S. (2010). SMEs and competitive advantage: A mix of innovation, marketing and ICT—The case of "made in Italy". In M. Russ (Ed.), *Knowledge management strategies for business development* (pp. 310–323). Hershey, PA: IGI Global. doi:10.4018/978-1-60566-348-7.ch014

Mariano, S., & Simionato, N. (2010). Where are we looking? A practical approach to managing knowledge captured from eye-tracking experiments: The experience of gulf air. In E. O'Brien, S. Clifford, & M. Southern (Eds.), *Knowledge management for process, organizational and marketing innovation: Tools and methods* (pp. 216–227). Hershey, PA: IGI Global. doi:10.4018/978-1-61520-829-6.ch013

Marques, M. B. (2014). The value of information and information services in knowledge society. In G. Jamil, A. Malheiro, & F. Ribeiro (Eds.), *Rethinking the conceptual base for new practical applications in information value and quality* (pp. 134–161). Hershey, PA: IGI Global. doi:10.4018/978-1-4666-4562-2.ch007

Masrom, M., Mahmood, N. H., & Al-Araimi, A. A. (2014). Exploring knowledge types and knowledge protection in organizations. In M. Chilton & J. Bloodgood (Eds.), *Knowledge management and competitive advantage: Issues and potential solutions* (pp. 271–280). Hershey, PA: IGI Global. doi:10.4018/978-1-4666-4679-7.ch016

Masterson, F. (2013). Knowledge management in practice: Using wikis to facilitate project-based learning. In S. Buckley & M. Jakovljevic (Eds.), *Knowledge management innovations for interdisciplinary education: Organizational applications* (pp. 385–401). Hershey, PA: IGI Global. doi:10.4018/978-1-4666-1969-2.ch019

Mattmann, C. A., Hart, A., Cinquini, L., Lazio, J., Khudikyan, S., & Jones, D. … Robnett, J. (2014). Scalable data mining, archiving, and big data management for the next generation astronomical telescopes. In W. Hu, & N. Kaabouch (Eds.), Big data management, technologies, and applications (pp. 196-221). Hershey, PA: IGI Global. doi:10.4018/978-1-4666-4699-5.ch009

Maule, R. W. (2011). Military knowledge management. In D. Schwartz & D. Te'eni (Eds.), *Encyclopedia of knowledge management* (2nd ed.; pp. 1125–1135). Hershey, PA: IGI Global. doi:10.4018/978-1-59904-931-1.ch108

Mavridis, I. (2011). Deploying privacy improved RBAC in web information systems. *International Journal of Information Technologies and Systems Approach*, *4*(2), 70–87. doi:10.4018/jitsa.2011070105

Mavridis, I. (2012). Deploying privacy improved RBAC in web information systems. In F. Stowell (Ed.), *Systems approach applications for developments in information technology* (pp. 298–315). Hershey, PA: IGI Global. doi:10.4018/978-1-4666-1562-5.ch020

McLaughlin, S. (2011). Assessing the impact of knowledge transfer mechanisms on supply chain performance. In M. Jennex (Ed.), *Global aspects and cultural perspectives on knowledge management: Emerging dimensions* (pp. 157–171). Hershey, PA: IGI Global. doi:10.4018/978-1-60960-555-1.ch011

Medina, J. M., & Spinola, M. D. (2011). Understanding the behavior of knowledge management pathways: The case of small manufacturers of footwear in Peru and Brazil. In M. Al-Shammari (Ed.), *Knowledge management in emerging economies: Social, organizational and cultural implementation* (pp. 261–271). Hershey, PA: IGI Global. doi:10.4018/978-1-61692-886-5.ch016

Meloche, J. A., Hasan, H., Willis, D., Pfaff, C. C., & Qi, Y. (2011). Cocreating corporate knowledge with a wiki. In M. Jennex (Ed.), *Global aspects and cultural perspectives on knowledge management: Emerging dimensions* (pp. 126–143). Hershey, PA: IGI Global. doi:10.4018/978-1-60960-555-1.ch009

Melzer, S. (2013). On the relationship between ontology-based and holistic representations in a knowledge management system. In M. Nazir Ahmad, R. Colomb, & M. Abdullah (Eds.), *Ontology-based applications for enterprise systems and knowledge management* (pp. 292–323). Hershey, PA: IGI Global. doi:10.4018/978-1-4666-1993-7.ch017

Mendes, E., & Baker, S. (2013). Using knowledge management and aggregation techniques to improve web effort estimation. In S. Saeed & I. Alsmadi (Eds.), *Knowledge-based processes in software development* (pp. 64–85). Hershey, PA: IGI Global. doi:10.4018/978-1-4666-4229-4.ch005

Metaxiotis, K. (2011). Healthcare knowledge management. In D. Schwartz & D. Te'eni (Eds.), *Encyclopedia of knowledge management* (2nd ed.; pp. 366–375). Hershey, PA: IGI Global. doi:10.4018/978-1-59904-931-1.ch035

Mikolajuk, Z. (2013). Community-based development of knowledge products. In M. Jennex (Ed.), *Dynamic models for knowledge-driven organizations* (pp. 268–281). Hershey, PA: IGI Global. doi:10.4018/978-1-4666-2485-6.ch016

Mischo, W. H., Schlembach, M. C., Bishoff, J., & German, E. M. (2012). User search activities within an academic library gateway: Implications for web-scale discovery systems. In M. Popp & D. Dallis (Eds.), *Planning and implementing resource discovery tools in academic libraries* (pp. 153–173). Hershey, PA: IGI Global. doi:10.4018/978-1-4666-1821-3.ch010

Mishra, B., & Shukla, K. K. (2014). Data mining techniques for software quality prediction. In *Software design and development: Concepts, methodologies, tools, and applications* (pp. 401–428). Hershey, PA: IGI Global. doi:10.4018/978-1-4666-4301-7.ch021

Moffett, S., Walker, T., & McAdam, R. (2014). Best value and performance management inspired change within UK councils: A knowledge management perspective. In Y. Al-Bastaki & A. Shajera (Eds.), *Building a competitive public sector with knowledge management strategy* (pp. 199–226). Hershey, PA: IGI Global. doi:10.4018/978-1-4666-4434-2.ch009

Mueller, C. E., & Bradley, K. D. (2011). Utilizing the Rasch model to develop and evaluate items for the tacit knowledge inventory for superintendents (TKIS). In M. Jennex (Ed.), *Global aspects and cultural perspectives on knowledge management: Emerging dimensions* (pp. 264–284). Hershey, PA: IGI Global. doi:10.4018/978-1-60960-555-1.ch017

Muhammed, S., Doll, W. J., & Deng, X. (2011). Impact of knowledge management practices on task knowledge: An individual level study. *International Journal of Knowledge Management, 7*(4), 1–21. doi:10.4018/jkm.2011100101

Muhammed, S., Doll, W. J., & Deng, X. (2011). Measuring knowledge management outcomes at the individual level: Towards a tool for research on organizational culture. In M. Jennex (Ed.), *Global aspects and cultural perspectives on knowledge management: Emerging dimensions* (pp. 1–18). Hershey, PA: IGI Global. doi:10.4018/978-1-60960-555-1.ch001

Muhammed, S., Doll, W. J., & Deng, X. (2013). Impact of knowledge management practices on task knowledge: An individual level study. In M. Jennex (Ed.), *Dynamic models for knowledge-driven organizations* (pp. 282–301). Hershey, PA: IGI Global. doi:10.4018/978-1-4666-2485-6.ch017

Murata, K. (2011). Knowledge creation and sharing in Japanese organisations: A socio-cultural perspective on ba. In G. Morais da Costa (Ed.), *Ethical issues and social dilemmas in knowledge management: Organizational innovation* (pp. 1–16). Hershey, PA: IGI Global. doi:10.4018/978-1-61520-873-9.ch001

Murphy, P. (2013). Systems of communication: Information, explanation, and imagination. In G. Yang (Ed.), *Multidisciplinary studies in knowledge and systems science* (pp. 63–78). Hershey, PA: IGI Global. doi:10.4018/978-1-4666-3998-0.ch006

Nach, H. (2013). Structuring knowledge for enterprise resource planning implementation through an ontology. In M. Nazir Ahmad, R. Colomb, & M. Abdullah (Eds.), *Ontology-based applications for enterprise systems and knowledge management* (pp. 25–42). Hershey, PA: IGI Global. doi:10.4018/978-1-4666-1993-7.ch002

Nah, F. F., Hong, W., Chen, L., & Lee, H. (2010). Information search patterns in e-commerce product comparison services. *Journal of Database Management, 21*(2), 26–40. doi:10.4018/jdm.2010040102

Nah, F. F., Hong, W., Chen, L., & Lee, H. (2012). Information search patterns in e-commerce product comparison services. In K. Siau (Ed.), *Cross-disciplinary models and applications of database management: Advancing approaches* (pp. 131–145). Hershey, PA: IGI Global. doi:10.4018/978-1-61350-471-0.ch006

Natarajan, R., & Shekar, B. (2011). Knowledge patterns in databases. In D. Schwartz & D. Te'eni (Eds.), *Encyclopedia of knowledge management* (2nd ed.; pp. 842–852). Hershey, PA: IGI Global. doi:10.4018/978-1-59904-931-1.ch081

Nelson, R. E., & Hsu, H. S. (2011). A social network perspective on knowledge management. In D. Schwartz & D. Te'eni (Eds.), *Encyclopedia of knowledge management* (2nd ed.; pp. 1470–1478). Hershey, PA: IGI Global. doi:10.4018/978-1-59904-931-1.ch140

Neto, R. C., & Souza, R. R. (2010). Knowledge management as an organizational process: From a theoretical framework to implementation guidelines. In E. O'Brien, S. Clifford, & M. Southern (Eds.), *Knowledge management for process, organizational and marketing innovation: Tools and methods* (pp. 16–35). Hershey, PA: IGI Global. doi:10.4018/978-1-61520-829-6.ch002

Newell, S. (2011). Understanding innovation processes. In D. Schwartz & D. Te'eni (Eds.), *Encyclopedia of knowledge management* (2nd ed.; pp. 1525–1535). Hershey, PA: IGI Global. doi:10.4018/978-1-59904-931-1.ch145

Nikabadi, M. S., & Zamanloo, S. (2012). A multidimensional structure for describing the influence of supply chain strategies, business strategies, and knowledge management strategies on knowledge sharing in supply chain. *International Journal of Knowledge Management, 8*(4), 50–70. doi:10.4018/jkm.2012100103

Nissen, M. (2014). Cyberspace and cloud knowledge. In *Harnessing dynamic knowledge principles in the technology-driven world* (pp. 193–204). Hershey, PA: IGI Global. doi:10.4018/978-1-4666-4727-5.ch012

Nissen, M. (2014). Social media knowledge. In *Harnessing dynamic knowledge principles in the technology-driven world* (pp. 219–227). Hershey, PA: IGI Global. doi:10.4018/978-1-4666-4727-5.ch014

Nissen, M. E. (2014). Harnessing knowledge power for competitive advantage. In M. Chilton & J. Bloodgood (Eds.), *Knowledge management and competitive advantage: Issues and potential solutions* (pp. 20–34). Hershey, PA: IGI Global. doi:10.4018/978-1-4666-4679-7.ch002

Nissen, M. E., & Levitt, R. E. (2011). Knowledge management research through computational experimentation. In D. Schwartz & D. Te'eni (Eds.), *Encyclopedia of knowledge management* (2nd ed.; pp. 728–737). Hershey, PA: IGI Global. doi:10.4018/978-1-59904-931-1.ch071

Nisula, A. (2014). Developing organizational renewal capability in the municipal (city) organization. In Y. Al-Bastaki & A. Shajera (Eds.), *Building a competitive public sector with knowledge management strategy* (pp. 151–172). Hershey, PA: IGI Global. doi:10.4018/978-1-4666-4434-2.ch007

Niu, B., Martin, P., & Powley, W. (2011). Towards autonomic workload management in DBMSs. In K. Siau (Ed.), *Theoretical and practical advances in information systems development: Emerging trends and approaches* (pp. 154–173). Hershey, PA: IGI Global. doi:10.4018/978-1-60960-521-6.ch008

Nobre, F. S., & Walker, D. S. (2011). A dynamic ability-based view of the organization. *International Journal of Knowledge Management*, *7*(2), 86–101. doi:10.4018/jkm.2011040105

O'Brien, J. (2014). Lessons from the private sector: A framework to be adopted in the public sector. In Y. Al-Bastaki & A. Shajera (Eds.), *Building a competitive public sector with knowledge management strategy* (pp. 173–198). Hershey, PA: IGI Global. doi:10.4018/978-1-4666-4434-2.ch008

Omari, A. (2013). Supporting companies management and improving their productivity through mining customers transactions. In *Data mining: Concepts, methodologies, tools, and applications* (pp. 1519–1533). Hershey, PA: IGI Global. doi:10.4018/978-1-4666-2455-9.ch079

Onwubiko, C. (2014). Modelling situation awareness information and system requirements for the mission using goal-oriented task analysis approach. In *Software design and development: Concepts, methodologies, tools, and applications* (pp. 460–478). Hershey, PA: IGI Global. doi:10.4018/978-1-4666-4301-7.ch023

Orth, A., Smolnik, S., & Jennex, M. E. (2011). The relevance of integration for knowledge management success: Towards conceptual and empirical evidence. In M. Jennex & S. Smolnik (Eds.), *Strategies for knowledge management success: Exploring organizational efficacy* (pp. 238–261). Hershey, PA: IGI Global. doi:10.4018/978-1-60566-709-6.ch013

Othman, A. K., & Abdullah, H. S. (2011). The influence of emotional intelligence on tacit knowledge sharing in service organizations. In M. Al-Shammari (Ed.), *Knowledge management in emerging economies: Social, organizational and cultural implementation* (pp. 171–185). Hershey, PA: IGI Global. doi:10.4018/978-1-61692-886-5.ch010

Pagallo, U. (2011). The trouble with digital copies: A short KM phenomenology. In G. Morais da Costa (Ed.), *Ethical issues and social dilemmas in knowledge management: Organizational innovation* (pp. 97–112). Hershey, PA: IGI Global. doi:10.4018/978-1-61520-873-9.ch007

Palte, R., Hertlein, M., Smolnik, S., & Riempp, G. (2013). The effects of a KM strategy on KM performance in professional services firms. In M. Jennex (Ed.), *Dynamic models for knowledge-driven organizations* (pp. 16–35). Hershey, PA: IGI Global. doi:10.4018/978-1-4666-2485-6.ch002

Pańkowska, M. (2014). Information technology prosumption acceptance by business information system consultants. In M. Pańkowska (Ed.), *Frameworks of IT prosumption for business development* (pp. 119–141). Hershey, PA: IGI Global. doi:10.4018/978-1-4666-4313-0.ch009

Pankowski, T. (2011). Pattern-based schema mapping and query answering in peer-to-peer XML data integration system. In L. Yan & Z. Ma (Eds.), *Advanced database query systems: Techniques, applications and technologies* (pp. 221–246). Hershey, PA: IGI Global. doi:10.4018/978-1-60960-475-2.ch009

Papoutsakis, H. (2010). New product development based on knowledge creation and technology education. In E. O'Brien, S. Clifford, & M. Southern (Eds.), *Knowledge management for process, organizational and marketing innovation: Tools and methods* (pp. 148–163). Hershey, PA: IGI Global. doi:10.4018/978-1-61520-829-6.ch009

Paquette, S. (2011). Applying knowledge management in the environmental and climate change sciences. In D. Schwartz & D. Te'eni (Eds.), *Encyclopedia of knowledge management* (2nd ed.; pp. 20–26). Hershey, PA: IGI Global. doi:10.4018/978-1-59904-931-1.ch003

Paquette, S. (2011). Customer knowledge management. In D. Schwartz & D. Te'eni (Eds.), *Encyclopedia of knowledge management* (2nd ed.; pp. 175–184). Hershey, PA: IGI Global. doi:10.4018/978-1-59904-931-1.ch017

Parker, K. R., & Nitse, P. S. (2011). Competitive intelligence gathering. In D. Schwartz & D. Te'eni (Eds.), *Encyclopedia of knowledge management* (2nd ed.; pp. 103–111). Hershey, PA: IGI Global. doi:10.4018/978-1-59904-931-1.ch011

Páscoa, C., & Tribolet, J. (2014). Maintaining organizational viability and performance: The organizational configuration map. In G. Jamil, A. Malheiro, & F. Ribeiro (Eds.), *Rethinking the conceptual base for new practical applications in information value and quality* (pp. 266–283). Hershey, PA: IGI Global. doi:10.4018/978-1-4666-4562-2.ch012

Paukert, M., Niederée, C., & Hemmje, M. (2011). Knowledge in innovation processes. In D. Schwartz & D. Te'eni (Eds.), *Encyclopedia of knowledge management* (2nd ed.; pp. 570–580). Hershey, PA: IGI Global. doi:10.4018/978-1-59904-931-1.ch055

Pawlak, P. (2011). Global "knowledge management" in humanist perspective. In G. Morais da Costa (Ed.), Ethical issues and social dilemmas in knowledge management: Organizational innovation (pp. 45-62). Hershey, PA: IGI Global. doi:10.4018/978-1-61520-873-9.ch004

Perry, M. (2013). Strategic knowledge management: A university application. In S. Buckley & M. Jakovljevic (Eds.), *Knowledge management innovations for interdisciplinary education: Organizational applications* (pp. 132–144). Hershey, PA: IGI Global. doi:10.4018/978-1-4666-1969-2.ch007

Pessoa, C. R., Silva, U. P., & Cruz, C. H. (2014). Information management in industrial areas: A knowledge management view. In G. Jamil, A. Malheiro, & F. Ribeiro (Eds.), *Rethinking the conceptual base for new practical applications in information value and quality* (pp. 378–395). Hershey, PA: IGI Global. doi:10.4018/978-1-4666-4562-2.ch017

Peter, H., & Greenidge, C. (2011). An ontology-based extraction framework for a semantic web application. *International Journal of Knowledge-Based Organizations, 1*(3), 56–71. doi:10.4018/ijkbo.2011070104

Peter, H., & Greenidge, C. (2013). An ontology-based extraction framework for a semantic web application. In J. Wang (Ed.), *Intelligence methods and systems advancements for knowledge-based business* (pp. 231–246). Hershey, PA: IGI Global. doi:10.4018/978-1-4666-1873-2.ch013

Pham, Q. T., & Hara, Y. (2011). KM approach for improving the labor productivity of Vietnamese enterprise. *International Journal of Knowledge Management, 7*(3), 27–42. doi:10.4018/jkm.2011070103

Pham, Q. T., & Hara, Y. (2013). KM approach for improving the labor productivity of Vietnamese enterprise. In M. Jennex (Ed.), *Dynamic models for knowledge-driven organizations* (pp. 206–219). Hershey, PA: IGI Global. doi:10.4018/978-1-4666-2485-6.ch012

Philpott, E., & Beaumont-Kerridge, J. (2010). Overcoming reticence to aid knowledge creation between universities and business: A case reviewed. In D. Harorimana (Ed.), *Cultural implications of knowledge sharing, management and transfer: Identifying competitive advantage* (pp. 355–368). Hershey, PA: IGI Global. doi:10.4018/978-1-60566-790-4.ch016

Pike, S., & Roos, G. (2011). Measuring and valuing knowledge-based intangible assets: Real business uses. In B. Vallejo-Alonso, A. Rodriguez-Castellanos, & G. Arregui-Ayastuy (Eds.), *Identifying, measuring, and valuing knowledge-based intangible assets: New perspectives* (pp. 268–293). Hershey, PA: IGI Global. doi:10.4018/978-1-60960-054-9.ch013

Pineda, J. L., Zapata, L. E., & Ramírez, J. (2010). Strengthening knowledge transfer between the university and enterprise: A conceptual model for collaboration. In D. Harorimana (Ed.), *Cultural implications of knowledge sharing, management and transfer: Identifying competitive advantage* (pp. 134–151). Hershey, PA: IGI Global. doi:10.4018/978-1-60566-790-4.ch007

Platonov, V., & Bergman, J. (2013). Cross-border cooperative network in the perspective of innovation dynamics. In J. Wang (Ed.), *Intelligence methods and systems advancements for knowledge-based business* (pp. 150–169). Hershey, PA: IGI Global. doi:10.4018/978-1-4666-1873-2.ch009

Poels, G. (2013). Understanding business domain models: The effect of recognizing resource-event-agent conceptual modeling structures. In K. Siau (Ed.), *Innovations in database design, web applications, and information systems management* (pp. 72–106). Hershey, PA: IGI Global. doi:10.4018/978-1-4666-2044-5.ch004

Poels, G., Decreus, K., Roelens, B., & Snoeck, M. (2013). Investigating goal-oriented requirements engineering for business processes. *Journal of Database Management, 24*(2), 35–71. doi:10.4018/jdm.2013040103

Pomares-Quimbaya, A., & Torres-Moreno, M. E. (2013). Knowledge management processes supported by ontology technologies. In M. Nazir Ahmad, R. Colomb, & M. Abdullah (Eds.), *Ontology-based applications for enterprise systems and knowledge management* (pp. 125–140). Hershey, PA: IGI Global. doi:10.4018/978-1-4666-1993-7.ch007

Ponis, S. T., Vagenas, G., & Koronis, E. (2010). Exploring the knowledge management landscape: A critical review of existing knowledge management frameworks. In D. Harorimana (Ed.), *Cultural implications of knowledge sharing, management and transfer: Identifying competitive advantage* (pp. 1–25). Hershey, PA: IGI Global. doi:10.4018/978-1-60566-790-4.ch001

Powers, S. M., & Salmon, C. (2010). Management of learning space. In D. Wu (Ed.), *Temporal structures in individual time management: Practices to enhance calendar tool design* (pp. 210–219). Hershey, PA: IGI Global. doi:10.4018/978-1-60566-776-8.ch015

Pretorius, A. B., & Coetzee, F. P. (2011). Model of a knowledge management support system for choosing intellectual capital assessment methods. In B. Vallejo-Alonso, A. Rodriguez-Castellanos, & G. Arregui-Ayastuy (Eds.), *Identifying, measuring, and valuing knowledge-based intangible assets: New perspectives* (pp. 336–359). Hershey, PA: IGI Global. doi:10.4018/978-1-60960-054-9.ch016

Pullinger, D. (2011). Mobilizing knowledge in the UK public sector: Current issues and discourse. In G. Morais da Costa (Ed.), *Ethical issues and social dilemmas in knowledge management: Organizational innovation* (pp. 232–249). Hershey, PA: IGI Global. doi:10.4018/978-1-61520-873-9.ch014

Rabaey, M. (2013). Complex adaptive systems thinking approach for intelligence base in support of intellectual capital management. In P. Ordóñez de Pablos, R. Tennyson, & J. Zhao (Eds.), *Intellectual capital strategy management for knowledge-based organizations* (pp. 122–141). Hershey, PA: IGI Global. doi:10.4018/978-1-4666-3655-2.ch007

Rabaey, M., & Mercken, R. (2013). Framework of knowledge and intelligence base: From intelligence to service. In *Data mining: Concepts, methodologies, tools, and applications* (pp. 474–502). Hershey, PA: IGI Global. doi:10.4018/978-1-4666-2455-9.ch023

Radziwill, N. M., & DuPlain, R. F. (2010). Quality and continuous improvement in knowledge management. In M. Russ (Ed.), *Knowledge management strategies for business development* (pp. 353–363). Hershey, PA: IGI Global. doi:10.4018/978-1-60566-348-7.ch017

Rahman, B. A., Saad, N. M., & Harun, M. S. (2010). Knowledge management orientation and business performance: The Malaysian manufacturing and service industries perspective. In K. Metaxiotis, F. Carrillo, & T. Yigitcanlar (Eds.), *Knowledge-based development for cities and societies: Integrated multi-level approaches* (pp. 315–328). Hershey, PA: IGI Global. doi:10.4018/978-1-61520-721-3.ch019

Randles, T. J., Blades, C. D., & Fadlalla, A. (2012). The knowledge spectrum. *International Journal of Knowledge Management*, *8*(2), 65–78. doi:10.4018/jkm.2012040104

Real, J. C., Leal, A., & Roldan, J. L. (2011). Measuring organizational learning as a multidimensional construct. In D. Schwartz & D. Te'eni (Eds.), *Encyclopedia of knowledge management* (2nd ed.; pp. 1101–1109). Hershey, PA: IGI Global. doi:10.4018/978-1-59904-931-1.ch105

Rech, J., & Bogner, C. (2010). Qualitative analysis of semantically enabled knowledge management systems in agile software engineering. *International Journal of Knowledge Management*, *6*(2), 66–85. doi:10.4018/jkm.2010040104

Rech, J., & Bogner, C. (2012). Qualitative analysis of semantically enabled knowledge management systems in agile software engineering. In M. Jennex (Ed.), *Conceptual models and outcomes of advancing knowledge management: New technologies* (pp. 144–164). Hershey, PA: IGI Global. doi:10.4018/978-1-4666-0035-5.ch008

Reis, R. S., & Curzi, Y. (2011). Knowledge integration in the creative process of globally distributed teams. In M. Al-Shammari (Ed.), *Knowledge management in emerging economies: Social, organizational and cultural implementation* (pp. 47–65). Hershey, PA: IGI Global. doi:10.4018/978-1-61692-886-5.ch003

Remli, M. A., & Deris, S. (2013). An approach for biological data integration and knowledge retrieval based on ontology, semantic web services composition, and AI planning. In M. Nazir Ahmad, R. Colomb, & M. Abdullah (Eds.), *Ontology-based applications for enterprise systems and knowledge management* (pp. 324–342). Hershey, PA: IGI Global. doi:10.4018/978-1-4666-1993-7.ch018

Reychav, I., Stein, E. W., Weisberg, J., & Glezer, C. (2012). The role of knowledge sharing in raising the task innovativeness of systems analysts. *International Journal of Knowledge Management, 8*(2), 1–22. doi:10.4018/jkm.2012040101

Reychav, I., & Weisberg, J. (2011). Human capital in knowledge creation, management, and utilization. In D. Schwartz & D. Te'eni (Eds.), *Encyclopedia of knowledge management* (2nd ed.; pp. 389–401). Hershey, PA: IGI Global. doi:10.4018/978-1-59904-931-1.ch037

Rhoads, E., O'Sullivan, K. J., & Stankosky, M. (2011). An evaluation of factors that influence the success of knowledge management practices in US federal agencies. In M. Jennex & S. Smolnik (Eds.), *Strategies for knowledge management success: Exploring organizational efficacy* (pp. 74–90). Hershey, PA: IGI Global. doi:10.4018/978-1-60566-709-6.ch005

Ribière, V. M. (2011). The effect of organizational trust on the success of codification and personalization KM approaches. In M. Jennex & S. Smolnik (Eds.), *Strategies for knowledge management success: Exploring organizational efficacy* (pp. 192–212). Hershey, PA: IGI Global. doi:10.4018/978-1-60566-709-6.ch011

Ribière, V. M., & Román, J. A. (2011). Knowledge flow. In D. Schwartz & D. Te'eni (Eds.), *Encyclopedia of knowledge management* (2nd ed.; pp. 549–559). Hershey, PA: IGI Global. doi:10.4018/978-1-59904-931-1.ch053

Ricceri, F., Guthrie, J., & Coyte, R. (2010). The management of knowledge resources within private organisations: Some European "better practice" illustrations. In E. O'Brien, S. Clifford, & M. Southern (Eds.), *Knowledge management for process, organizational and marketing innovation: Tools and methods* (pp. 36–61). Hershey, PA: IGI Global. doi:10.4018/978-1-61520-829-6.ch003

Riss, U. V. (2011). Pattern-based task management as means of organizational knowledge maturing. *International Journal of Knowledge-Based Organizations, 1*(1), 20–41. doi:10.4018/ijkbo.2011010102

Riss, U. V. (2013). Pattern-based task management as means of organizational knowledge maturing. In J. Wang (Ed.), *Intelligence methods and systems advancements for knowledge-based business* (pp. 1–23). Hershey, PA: IGI Global. doi:10.4018/978-1-4666-1873-2.ch001

Roos, G. (2013). The role of intellectual capital in business model innovation: An empirical study. In P. Ordóñez de Pablos, R. Tennyson, & J. Zhao (Eds.), *Intellectual capital strategy management for knowledge-based organizations* (pp. 76–121). Hershey, PA: IGI Global. doi:10.4018/978-1-4666-3655-2.ch006

Rothberg, H. N., & Klingenberg, B. (2010). Learning before doing: A theoretical perspective and practical lessons from a failed cross-border knowledge transfer initiative. In D. Harorimana (Ed.), *Cultural implications of knowledge sharing, management and transfer: Identifying competitive advantage* (pp. 277–294). Hershey, PA: IGI Global. doi:10.4018/978-1-60566-790-4.ch013

Ruano-Mayoral, M., Colomo-Palacios, R., García-Crespo, Á., & Gómez-Berbís, J. M. (2012). Software project managers under the team software process: A study of competences based on literature. In J. Wang (Ed.), *Project management techniques and innovations in information technology* (pp. 115–126). Hershey, PA: IGI Global. doi:10.4018/978-1-4666-0930-3.ch007

Russell, S. (2010). Knowledge management and project management in 3D: A virtual world extension. In E. O'Brien, S. Clifford, & M. Southern (Eds.), *Knowledge management for process, organizational and marketing innovation: Tools and methods* (pp. 62–78). Hershey, PA: IGI Global. doi:10.4018/978-1-61520-829-6.ch004

Ryan, G., & Shinnick, E. (2011). Knowledge and intellectual property rights: An economics perspective. In D. Schwartz & D. Te'eni (Eds.), *Encyclopedia of knowledge management* (2nd ed.; pp. 489–496). Hershey, PA: IGI Global. doi:10.4018/978-1-59904-931-1.ch047

Sabetzadeh, F., & Tsui, E. (2013). Delivering knowledge services in the cloud. In G. Yang (Ed.), *Multidisciplinary studies in knowledge and systems science* (pp. 247–254). Hershey, PA: IGI Global. doi:10.4018/978-1-4666-3998-0.ch017

Sáenz, J., & Aramburu, N. (2011). Organizational conditions as catalysts for successful people-focused knowledge sharing initiatives: An empirical study. *International Journal of Knowledge-Based Organizations, 1*(2), 39–56. doi:10.4018/ijkbo.2011040103

Sáenz, J., & Aramburu, N. (2013). Organizational conditions as catalysts for successful people-focused knowledge sharing initiatives: An empirical study. In J. Wang (Ed.), *Intelligence methods and systems advancements for knowledge-based business* (pp. 263–280). Hershey, PA: IGI Global. doi:10.4018/978-1-4666-1873-2.ch015

Sakr, S., & Al-Naymat, G. (2011). Relational techniques for storing and querying RDF data: An overview. In L. Yan & Z. Ma (Eds.), *Advanced database query systems: Techniques, applications and technologies* (pp. 269–285). Hershey, PA: IGI Global. doi:10.4018/978-1-60960-475-2.ch011

Salem, P. J. (2013). The use of mixed methods in organizational communication research. In M. Bocarnea, R. Reynolds, & J. Baker (Eds.), *Online instruments, data collection, and electronic measurements: Organizational advancements* (pp. 24–39). Hershey, PA: IGI Global. doi:10.4018/978-1-4666-2172-5.ch002

Salisbury, M. (2011). A framework for managing the life cycle of knowledge in global organizations. In M. Jennex (Ed.), *Global aspects and cultural perspectives on knowledge management: Emerging dimensions* (pp. 64–80). Hershey, PA: IGI Global. doi:10.4018/978-1-60960-555-1.ch005

Salleh, K. (2014). Drivers, benefits, and challenges of knowledge management in electronic government: Preliminary examination. In Y. Al-Bastaki & A. Shajera (Eds.), *Building a competitive public sector with knowledge management strategy* (pp. 135–150). Hershey, PA: IGI Global. doi:10.4018/978-1-4666-4434-2.ch006

Salleh, K., Ikhsan, S. O., & Ahmad, S. N. (2011). Knowledge management enablers and knowledge sharing process: A case study of public sector accounting organization in Malaysia. In M. Al-Shammari (Ed.), *Knowledge management in emerging economies: Social, organizational and cultural implementation* (pp. 199–211). Hershey, PA: IGI Global. doi:10.4018/978-1-61692-886-5.ch012

Saunders, C. (2011). Knowledge sharing in legal practice. In D. Schwartz & D. Te'eni (Eds.), *Encyclopedia of knowledge management* (2nd ed.; pp. 935–945). Hershey, PA: IGI Global. doi:10.4018/978-1-59904-931-1.ch089

Scarso, E., & Bolisani, E. (2011). Knowledge intermediation. In D. Schwartz & D. Te'eni (Eds.), *Encyclopedia of knowledge management* (2nd ed.; pp. 601–611). Hershey, PA: IGI Global. doi:10.4018/978-1-59904-931-1.ch058

Scarso, E., & Bolisani, E. (2011). Managing professions for knowledge management. *International Journal of Knowledge Management, 7*(3), 61–75. doi:10.4018/jkm.2011070105

Scarso, E., & Bolisani, E. (2013). Managing professions for knowledge management. In M. Jennex (Ed.), *Dynamic models for knowledge-driven organizations* (pp. 238–253). Hershey, PA: IGI Global. doi:10.4018/978-1-4666-2485-6.ch014

Scarso, E., Bolisani, E., & Padova, A. (2011). The complex issue of measuring KM performance: Lessons from the practice. In B. Vallejo-Alonso, A. Rodriguez-Castellanos, & G. Arregui-Ayastuy (Eds.), *Identifying, measuring, and valuing knowledge-based intangible assets: New perspectives* (pp. 208–230). Hershey, PA: IGI Global. doi:10.4018/978-1-60960-054-9.ch010

Schumann, C., & Tittmann, C. (2010). Potentials for externalizing and measuring of tacit knowledge within knowledge nodes in the context of knowledge networks. In D. Harorimana (Ed.), *Cultural implications of knowledge sharing, management and transfer: Identifying competitive advantage* (pp. 84–107). Hershey, PA: IGI Global. doi:10.4018/978-1-60566-790-4.ch005

Schwartz, D. (2011). An Aristotelian view of knowledge for knowledge management. In D. Schwartz & D. Te'eni (Eds.), *Encyclopedia of knowledge management* (2nd ed.; pp. 39–48). Hershey, PA: IGI Global. doi:10.4018/978-1-59904-931-1.ch005

Senaratne, S., & Victoria, M. F. (2014). Building a supportive culture for sustained organisational learning in public sectors. In Y. Al-Bastaki & A. Shajera (Eds.), *Building a competitive public sector with knowledge management strategy* (pp. 118–134). Hershey, PA: IGI Global. doi:10.4018/978-1-4666-4434-2.ch005

Shah, A., Singhera, Z., & Ahsan, S. (2011). Web services for bioinformatics. In M. Al-Mutairi & L. Mohammed (Eds.), *Cases on ICT utilization, practice and solutions: Tools for managing day-to-day issues* (pp. 28–46). Hershey, PA: IGI Global. doi:10.4018/978-1-60960-015-0.ch003

Shajera, A., & Al-Bastaki, Y. (2014). Organisational readiness for knowledge management: Bahrain public sector case study. In Y. Al-Bastaki & A. Shajera (Eds.), *Building a competitive public sector with knowledge management strategy* (pp. 104–117). Hershey, PA: IGI Global. doi:10.4018/978-1-4666-4434-2.ch004

Sharma, A. K., Goswami, A., & Gupta, D. (2011). An extended relational model & SQL for fuzzy multidatabases. In L. Yan & Z. Ma (Eds.), *Advanced database query systems: Techniques, applications and technologies* (pp. 185–219). Hershey, PA: IGI Global. doi:10.4018/978-1-60960-475-2.ch008

Sharma, R., Banati, H., & Bedi, P. (2012). Building socially-aware e-learning systems through knowledge management. *International Journal of Knowledge Management*, *8*(3), 1–26. doi:10.4018/jkm.2012070101

Sharma, R. S., Chandrasekar, G., & Vaitheeswaran, B. (2012). A knowledge framework for development: Empirical investigation of 30 societies. In M. Jennex (Ed.), *Conceptual models and outcomes of advancing knowledge management: New technologies* (pp. 244–265). Hershey, PA: IGI Global. doi:10.4018/978-1-4666-0035-5.ch013

Shaw, D. (2011). Mapping group knowledge. In D. Schwartz & D. Te'eni (Eds.), *Encyclopedia of knowledge management* (2nd ed.; pp. 1072–1081). Hershey, PA: IGI Global. doi:10.4018/978-1-59904-931-1.ch102

Sheluhin, O. I., & Atayero, A. A. (2013). Principles of modeling in information communication systems and networks. In A. Atayero & O. Sheluhin (Eds.), *Integrated models for information communication systems and networks: Design and development* (pp. 1–15). Hershey, PA: IGI Global. doi:10.4018/978-1-4666-2208-1.ch001

Sheluhin, O. I., & Garmashev, A. V. (2013). Numerical methods of multifractal analysis in information communication systems and networks. In A. Atayero & O. Sheluhin (Eds.), *Integrated models for information communication systems and networks: Design and development* (pp. 16–46). Hershey, PA: IGI Global. doi:10.4018/978-1-4666-2208-1.ch002

Siau, K., Long, Y., & Ling, M. (2010). Toward a unified model of information systems development success. *Journal of Database Management*, *21*(1), 80–101. doi:10.4018/jdm.2010112304

Siau, K., Long, Y., & Ling, M. (2012). Toward a unified model of information systems development success. In K. Siau (Ed.), *Cross-disciplinary models and applications of database management: Advancing approaches* (pp. 80–102). Hershey, PA: IGI Global. doi:10.4018/978-1-61350-471-0.ch004

Simard, A. J., & Jourdeuil, P. (2014). Knowledge manageability: A new paradigm. In Y. Al-Bastaki & A. Shajera (Eds.), *Building a competitive public sector with knowledge management strategy* (pp. 1–52). Hershey, PA: IGI Global. doi:10.4018/978-1-4666-4434-2.ch001

Simonette, M. J., & Spina, E. (2014). Enabling IT innovation through soft systems engineering. In M. Pańkowska (Ed.), *Frameworks of IT prosumption for business development* (pp. 64–72). Hershey, PA: IGI Global. doi:10.4018/978-1-4666-4313-0.ch005

Sivaramakrishnan, S., Delbaere, M., Zhang, D., & Bruning, E. (2012). Critical success factors and outcomes of market knowledge management: A conceptual model and empirical evidence. In M. Jennex (Ed.), *Conceptual models and outcomes of advancing knowledge management: New technologies* (pp. 165–185). Hershey, PA: IGI Global. doi:10.4018/978-1-4666-0035-5.ch009

Small, C. T., & Sage, A. P. (2010). A complex adaptive systems-based enterprise knowledge sharing model. In D. Paradice (Ed.), *Emerging systems approaches in information technologies: Concepts, theories, and applications* (pp. 137–155). Hershey, PA: IGI Global. doi:10.4018/978-1-60566-976-2.ch009

Smedlund, A. (2011). Social network structures for explicit, tacit and potential knowledge. In M. Jennex (Ed.), *Global aspects and cultural perspectives on knowledge management: Emerging dimensions* (pp. 81–90). Hershey, PA: IGI Global. doi:10.4018/978-1-60960-555-1.ch006

Smith, A. D. (2013). Competitive uses of information and knowledge management tools: Case study of supplier-side management. *International Journal of Knowledge-Based Organizations*, *3*(1), 71–87. doi:10.4018/ijkbo.2013010105

Smith, P., & Coakes, E. (2011). Exploiting KM in support of innovation and change. In A. Eardley & L. Uden (Eds.), *Innovative knowledge management: Concepts for organizational creativity and collaborative design* (pp. 242–252). Hershey, PA: IGI Global. doi:10.4018/978-1-60566-701-0.ch015

Smith, T. A., Mills, A. M., & Dion, P. (2010). Linking business strategy and knowledge management capabilities for organizational effectiveness. *International Journal of Knowledge Management*, *6*(3), 22–43. doi:10.4018/jkm.2010070102

Smith, T. A., Mills, A. M., & Dion, P. (2012). Linking business strategy and knowledge management capabilities for organizational effectiveness. In M. Jennex (Ed.), *Conceptual models and outcomes of advancing knowledge management: New technologies* (pp. 186–207). Hershey, PA: IGI Global. doi:10.4018/978-1-4666-0035-5.ch010

Smuts, H., van der Merwe, A., & Loock, M. (2011). Key characteristics relevant for selecting knowledge management software tools. In A. Eardley & L. Uden (Eds.), *Innovative knowledge management: Concepts for organizational creativity and collaborative design* (pp. 18–39). Hershey, PA: IGI Global. doi:10.4018/978-1-60566-701-0.ch002

Soffer, P., & Kaner, M. (2013). Complementing business process verification by validity analysis: A theoretical and empirical evaluation. In K. Siau (Ed.), *Innovations in database design, web applications, and information systems management* (pp. 265–288). Hershey, PA: IGI Global. doi:10.4018/978-1-4666-2044-5.ch010

Soffer, P., Kaner, M., & Wand, Y. (2012). Assigning ontological meaning to workflow nets. In K. Siau (Ed.), *Cross-disciplinary models and applications of database management: Advancing approaches* (pp. 209–244). Hershey, PA: IGI Global. doi:10.4018/978-1-61350-471-0.ch009

Sohrabi, B., Raeesi, I., & Khanlari, A. (2010). Intellectual capital components, measurement and management: A literature survey of concepts and measures. In P. López Sáez, G. Castro, J. Navas López, & M. Delgado Verde (Eds.), *Intellectual capital and technological innovation: Knowledge-based theory and practice* (pp. 1–38). Hershey, PA: IGI Global. doi:10.4018/978-1-61520-875-3.ch001

Sohrabi, B., Raeesi, I., Khanlari, A., & Forouzandeh, S. (2011). A comprehensive model for assessing the organizational readiness of knowledge management. In M. Jennex (Ed.), *Global aspects and cultural perspectives on knowledge management: Emerging dimensions* (pp. 30–48). Hershey, PA: IGI Global. doi:10.4018/978-1-60960-555-1.ch003

Sparrow, J. (2011). Knowledge management in small and medium sized enterprises. In D. Schwartz & D. Te'eni (Eds.), *Encyclopedia of knowledge management* (2nd ed.; pp. 671–681). Hershey, PA: IGI Global. doi:10.4018/978-1-59904-931-1.ch065

Stam, C. D. (2011). Making sense of knowledge productivity. In B. Vallejo-Alonso, A. Rodriguez-Castellanos, & G. Arregui-Ayastuy (Eds.), *Identifying, measuring, and valuing knowledge-based intangible assets: New perspectives* (pp. 133–155). Hershey, PA: IGI Global. doi:10.4018/978-1-60960-054-9.ch007

Stamkopoulos, K., Pitoura, E., Vassiliadis, P., & Zarras, A. (2012). Accelerating web service workflow execution via intelligent allocation of services to servers. In K. Siau (Ed.), *Cross-disciplinary models and applications of database management: Advancing approaches* (pp. 385–416). Hershey, PA: IGI Global. doi:10.4018/978-1-61350-471-0.ch016

Sterling, L. (2011). Applying agents within knowledge management. In D. Schwartz & D. Te'eni (Eds.), *Encyclopedia of knowledge management* (2nd ed.; pp. 12–19). Hershey, PA: IGI Global. doi:10.4018/978-1-59904-931-1.ch002

Su, S., & Chiong, R. (2011). Business intelligence. In D. Schwartz & D. Te'eni (Eds.), *Encyclopedia of knowledge management* (2nd ed.; pp. 72–80). Hershey, PA: IGI Global. doi:10.4018/978-1-59904-931-1.ch008

Su, W. B., Li, X., & Chow, C. W. (2012). Exploring the extent and impediments of knowledge sharing in Chinese business enterprise. In M. Jennex (Ed.), *Conceptual models and outcomes of advancing knowledge management: New technologies* (pp. 266–290). Hershey, PA: IGI Global. doi:10.4018/978-1-4666-0035-5.ch014

Subramanian, D. V., & Geetha, A. (2012). Application of multi-dimensional metric model, database, and WAM for KM system evaluation. *International Journal of Knowledge Management, 8*(4), 1–21. doi:10.4018/jkm.2012100101

Surendran, A., & Samuel, P. (2013). Knowledge-based code clone approach in embedded and real-time systems. In S. Saeed & I. Alsmadi (Eds.), *Knowledge-based processes in software development* (pp. 49–62). Hershey, PA: IGI Global. doi:10.4018/978-1-4666-4229-4.ch004

Takahashi, Y. (2011). The importance of balancing knowledge protection and knowledge interchange. In G. Morais da Costa (Ed.), *Ethical issues and social dilemmas in knowledge management: Organizational innovation* (pp. 180–198). Hershey, PA: IGI Global. doi:10.4018/978-1-61520-873-9.ch011

Talet, A. N., Alhawari, S., & Alryalat, H. (2012). The outcome of knowledge process for customers of Jordanian companies on the achievement of customer knowledge retention. In M. Jennex (Ed.), *Conceptual models and outcomes of advancing knowledge management: New technologies* (pp. 45–61). Hershey, PA: IGI Global. doi:10.4018/978-1-4666-0035-5.ch003

Talet, A. N., Alhawari, S., Mansour, E., & Alryalat, H. (2011). The practice of Jordanian business to attain customer knowledge acquisition. *International Journal of Knowledge Management, 7*(2), 49–67. doi:10.4018/jkm.2011040103

Tanner, K. (2011). The role of emotional capital in organisational KM. In D. Schwartz & D. Te'eni (Eds.), *Encyclopedia of knowledge management* (2nd ed.; pp. 1396–1409). Hershey, PA: IGI Global. doi:10.4018/978-1-59904-931-1.ch133

Tauber, D., & Schwartz, D. G. (2011). Integrating knowledge management with the systems analysis process. In D. Schwartz & D. Te'eni (Eds.), *Encyclopedia of knowledge management* (2nd ed.; pp. 431–441). Hershey, PA: IGI Global. doi:10.4018/978-1-59904-931-1.ch041

Tavana, M., Busch, T. E., & Davis, E. L. (2011). Modeling operational robustness and resiliency with high-level Petri nets. *International Journal of Knowledge-Based Organizations*, *1*(2), 17–38. doi:10.4018/ijkbo.2011040102

Taxén, L. (2010). Aligning business and knowledge strategies: A practical approach for aligning business and knowledge strategies. In M. Russ (Ed.), *Knowledge management strategies for business development* (pp. 277–308). Hershey, PA: IGI Global. doi:10.4018/978-1-60566-348-7.ch013

Te'eni, D. (2011). Knowledge for communicating knowledge. In D. Schwartz & D. Te'eni (Eds.), *Encyclopedia of knowledge management* (2nd ed.; pp. 560–569). Hershey, PA: IGI Global. doi:10.4018/978-1-59904-931-1.ch054

Toiviainen, H., & Kerosuo, H. (2013). Development curriculum for knowledge-based organizations: Lessons from a learning network. *International Journal of Knowledge-Based Organizations*, *3*(3), 1–18. doi:10.4018/ijkbo.2013070101

Tran, B. (2014). The human element of the knowledge worker: Identifying, managing, and protecting the intellectual capital within knowledge management. In M. Chilton & J. Bloodgood (Eds.), *Knowledge management and competitive advantage: Issues and potential solutions* (pp. 281–303). Hershey, PA: IGI Global. doi:10.4018/978-1-4666-4679-7.ch017

Tsamoura, E., Gounaris, A., & Manolopoulos, Y. (2011). Optimal service ordering in decentralized queries over web services. *International Journal of Knowledge-Based Organizations*, *1*(2), 1–16. doi:10.4018/ijkbo.2011040101

Tsamoura, E., Gounaris, A., & Manolopoulos, Y. (2013). Optimal service ordering in decentralized queries over web services. In J. Wang (Ed.), *Intelligence methods and systems advancements for knowledge-based business* (pp. 43–58). Hershey, PA: IGI Global. doi:10.4018/978-1-4666-1873-2.ch003

Tull, J. (2013). Slow knowledge: The case for savouring learning and innovation. In S. Buckley & M. Jakovljevic (Eds.), *Knowledge management innovations for interdisciplinary education: Organizational applications* (pp. 274–297). Hershey, PA: IGI Global. doi:10.4018/978-1-4666-1969-2.ch014

Turner, G., & Minonne, C. (2013). Effective knowledge management through measurement. In S. Buckley & M. Jakovljevic (Eds.), *Knowledge management innovations for interdisciplinary education: Organizational applications* (pp. 145–176). Hershey, PA: IGI Global. doi:10.4018/978-1-4666-1969-2.ch008

Uden, L., & Eardley, A. (2011). Knowledge sharing in the learning process: Experience with problem-based learning. In A. Eardley & L. Uden (Eds.), *Innovative knowledge management: Concepts for organizational creativity and collaborative design* (pp. 215–229). Hershey, PA: IGI Global. doi:10.4018/978-1-60566-701-0.ch013

Upadhyaya, S., Rao, H. R., & Padmanabhan, G. (2011). Secure knowledge management. In D. Schwartz & D. Te'eni (Eds.), *Encyclopedia of knowledge management* (2nd ed.; pp. 1429–1437). Hershey, PA: IGI Global. doi:10.4018/978-1-59904-931-1.ch136

Urbancová, H., & Königová, M. (2013). The influence of the application of business continuity management, knowledge management, and knowledge continuity management on the innovation in organizations. In S. Buckley & M. Jakovljevic (Eds.), *Knowledge management innovations for interdisciplinary education: Organizational applications* (pp. 254–273). Hershey, PA: IGI Global. doi:10.4018/978-1-4666-1969-2.ch013

Van Canh, T., & Zyngier, S. (2014). Using ERG theory as a lens to understand the sharing of academic tacit knowledge: Problems and issues in developing countries – Perspectives from Vietnam. In M. Chilton & J. Bloodgood (Eds.), *Knowledge management and competitive advantage: Issues and potential solutions* (pp. 174–201). Hershey, PA: IGI Global. doi:10.4018/978-1-4666-4679-7.ch010

Vat, K. H. (2011). Knowledge synthesis framework. In D. Schwartz & D. Te'eni (Eds.), *Encyclopedia of knowledge management* (2nd ed.; pp. 955–966). Hershey, PA: IGI Global. doi:10.4018/978-1-59904-931-1.ch091

Vert, S. (2012). Extensions of web browsers useful to knowledge workers. In C. Jouis, I. Biskri, J. Ganascia, & M. Roux (Eds.), *Next generation search engines: Advanced models for information retrieval* (pp. 239–273). Hershey, PA: IGI Global. doi:10.4018/978-1-4666-0330-1.ch011

Wagner, L., & Van Belle, J. (2011). Web mining for strategic competitive intelligence: South African experiences and a practical methodology. In M. Al-Shammari (Ed.), *Knowledge management in emerging economies: Social, organizational and cultural implementation* (pp. 1–19). Hershey, PA: IGI Global. doi:10.4018/978-1-61692-886-5.ch001

Wautelet, Y., Schinckus, C., & Kolp, M. (2010). Towards knowledge evolution in software engineering: An epistemological approach. *International Journal of Information Technologies and Systems Approach*, *3*(1), 21–40. doi:10.4018/jitsa.2010100202

Weiß, S., Makolm, J., Ipsmiller, D., & Egger, N. (2011). DYONIPOS: Proactive knowledge supply. In M. Jennex & S. Smolnik (Eds.), *Strategies for knowledge management success: Exploring organizational efficacy* (pp. 277–287). Hershey, PA: IGI Global. doi:10.4018/978-1-60566-709-6.ch015

Welschen, J., Todorova, N., & Mills, A. M. (2012). An investigation of the impact of intrinsic motivation on organizational knowledge sharing. *International Journal of Knowledge Management*, *8*(2), 23–42. doi:10.4018/jkm.2012040102

Wickramasinghe, N. (2011). Knowledge creation. In D. Schwartz & D. Te'eni (Eds.), *Encyclopedia of knowledge management* (2nd ed.; pp. 527–538). Hershey, PA: IGI Global. doi:10.4018/978-1-59904-931-1.ch051

Wickramasinghe, N. (2011). Knowledge management: The key to delivering superior healthcare solutions. In A. Eardley & L. Uden (Eds.), *Innovative knowledge management: Concepts for organizational creativity and collaborative design* (pp. 190–203). Hershey, PA: IGI Global. doi:10.4018/978-1-60566-701-0.ch011

Wijnhoven, F. (2011). Operational knowledge management. In D. Schwartz & D. Te'eni (Eds.), *Encyclopedia of knowledge management* (2nd ed.; pp. 1237–1249). Hershey, PA: IGI Global. doi:10.4018/978-1-59904-931-1.ch118

Williams, R. (2011). A Knowledge Process Cycle. In D. Schwartz & D. Te'eni (Eds.), *Encyclopedia of knowledge management* (2nd ed.; pp. 853–866). Hershey, PA: IGI Global. doi:10.4018/978-1-59904-931-1.ch082

Wilson, R. L., Rosen, P. A., & Al-Ahmadi, M. S. (2011). Knowledge structure and data mining techniques. In D. Schwartz & D. Te'eni (Eds.), *Encyclopedia of knowledge management* (2nd ed.; pp. 946–954). Hershey, PA: IGI Global. doi:10.4018/978-1-59904-931-1.ch090

Wimmer, H., Yoon, V., & Rada, R. (2013). Integrating knowledge sources: An ontological approach. *International Journal of Knowledge Management*, *9*(1), 60–75. doi:10.4018/jkm.2013010104

Woods, S., Poteet, S. R., Kao, A., & Quach, L. (2011). Knowledge dissemination in portals. In D. Schwartz & D. Te'eni (Eds.), *Encyclopedia of knowledge management* (2nd ed.; pp. 539–548). Hershey, PA: IGI Global. doi:10.4018/978-1-59904-931-1.ch052

Worden, D. (2010). Agile alignment of enterprise execution capabilities with strategy. In M. Russ (Ed.), *Knowledge management strategies for business development* (pp. 45–62). Hershey, PA: IGI Global. doi:10.4018/978-1-60566-348-7.ch003

Wu, D. (2010). Who are effective time managers? Bivariate correlation analysis and hypotheses testing. In D. Wu (Ed.), *Temporal structures in individual time management: Practices to enhance calendar tool design* (pp. 116–138). Hershey, PA: IGI Global. doi:10.4018/978-1-60566-776-8.ch009

Wu, J., Du, H., Li, X., & Li, P. (2010). Creating and delivering a successful knowledge management strategy. In M. Russ (Ed.), *Knowledge management strategies for business development* (pp. 261–276). Hershey, PA: IGI Global. doi:10.4018/978-1-60566-348-7.ch012

Wu, J., Liu, N., & Xuan, Z. (2013). Simulation on knowledge transfer processes from the perspectives of individual's mentality and behavior. In G. Yang (Ed.), *Multidisciplinary studies in knowledge and systems science* (pp. 233–246). Hershey, PA: IGI Global. doi:10.4018/978-1-4666-3998-0.ch016

Wu, J., Wang, S., & Pan, D. (2013). Evaluation of technological influence power of enterprises through the enterprise citation network. In G. Yang (Ed.), *Multidisciplinary studies in knowledge and systems science* (pp. 34–44). Hershey, PA: IGI Global. doi:10.4018/978-1-4666-3998-0.ch003

Wu, S. T. (2011). Innovation in new technology and knowledge management: Comparative case studies of its evolution during a quarter century of change. In A. Eardley & L. Uden (Eds.), *Innovative knowledge management: Concepts for organizational creativity and collaborative design* (pp. 77–93). Hershey, PA: IGI Global. doi:10.4018/978-1-60566-701-0.ch005

Xiao, L., & Pei, Y. (2013). A task context aware physical distribution knowledge service system. In G. Yang (Ed.), *Multidisciplinary studies in knowledge and systems science* (pp. 18–33). Hershey, PA: IGI Global. doi:10.4018/978-1-4666-3998-0.ch002

Xu, D., & Wang, H. (2011). Integration of knowledge management and e-learning. In D. Schwartz & D. Te'eni (Eds.), *Encyclopedia of knowledge management* (2nd ed.; pp. 442–451). Hershey, PA: IGI Global. doi:10.4018/978-1-59904-931-1.ch042

Yaniv, E., & Schwartz, D. G. (2011). Organizational attention. In D. Schwartz & D. Te'eni (Eds.), *Encyclopedia of knowledge management* (2nd ed.; pp. 1270–1279). Hershey, PA: IGI Global. doi:10.4018/978-1-59904-931-1.ch121

Yeung, C. L., Cheung, C. F., Wang, W. M., & Tsui, E. (2013). A study of organizational narrative simulation for decision support. In G. Yang (Ed.), *Multidisciplinary studies in knowledge and systems science* (pp. 179–192). Hershey, PA: IGI Global. doi:10.4018/978-1-4666-3998-0.ch013

Yigitcanlar, T. (2013). Moving towards a knowledge city? Brisbane's experience in knowledge-based urban development. In J. Wang (Ed.), *Intelligence methods and systems advancements for knowledge-based business* (pp. 114–131). Hershey, PA: IGI Global. doi:10.4018/978-1-4666-1873-2.ch007

Yigitcanlar, T., & Martinez-Fernandez, C. (2010). Making space and place for knowledge production: Socio-spatial development of knowledge community precincts. In K. Metaxiotis, F. Carrillo, & T. Yigitcanlar (Eds.), *Knowledge-based development for cities and societies: Integrated multi-level approaches* (pp. 99–117). Hershey, PA: IGI Global. doi:10.4018/978-1-61520-721-3.ch006

Yıldırım, A. A., Özdoğan, C., & Watson, D. (2014). Parallel data reduction techniques for big datasets. In W. Hu & N. Kaabouch (Eds.), *Big data management, technologies, and applications* (pp. 72–93). Hershey, PA: IGI Global. doi:10.4018/978-1-4666-4699-5.ch004

Yoon, K. S. (2012). Measuring the influence of expertise and epistemic engagement to the practice of knowledge management. *International Journal of Knowledge Management, 8*(1), 40–70. doi:10.4018/jkm.2012010103

Yusof, Z. M., & Ismail, M. B. (2011). Factors affecting knowledge sharing practice in Malaysia: A preliminary overview. In M. Al-Shammari (Ed.), *Knowledge management in emerging economies: Social, organizational and cultural implementation* (pp. 157–170). Hershey, PA: IGI Global. doi:10.4018/978-1-61692-886-5.ch009

Zapata-Cantú, L., Ramírez, J., & Pineda, J. L. (2011). HRM adaptation to knowledge management initiatives: Three Mexican cases. In M. Al-Shammari (Ed.), *Knowledge management in emerging economies: Social, organizational and cultural implementation* (pp. 273–293). Hershey, PA: IGI Global. doi:10.4018/978-1-61692-886-5.ch017

Zarri, G. P. (2011). Knowledge representation. In D. Schwartz & D. Te'eni (Eds.), *Encyclopedia of knowledge management* (2nd ed.; pp. 878–892). Hershey, PA: IGI Global. doi:10.4018/978-1-59904-931-1.ch084

Zarri, G. P. (2011). RDF and OWL for knowledge management. In D. Schwartz & D. Te'eni (Eds.), *Encyclopedia of knowledge management* (2nd ed.; pp. 1355–1373). Hershey, PA: IGI Global. doi:10.4018/978-1-59904-931-1.ch130

Zhang, Y., Wang, Y., Colucci, W., & Wang, Z. (2013). The paradigm shift in organizational research. In J. Wang (Ed.), *Intelligence methods and systems advancements for knowledge-based business* (pp. 60–74). Hershey, PA: IGI Global. doi:10.4018/978-1-4666-1873-2.ch004

Zhang, Z. J. (2011). Managing customer knowledge with social software. In D. Schwartz & D. Te'eni (Eds.), *Encyclopedia of knowledge management* (2nd ed.; pp. 1046–1053). Hershey, PA: IGI Global. doi:10.4018/978-1-59904-931-1.ch099

Zyngier, S. (2011). Governance of knowledge management. In D. Schwartz & D. Te'eni (Eds.), *Encyclopedia of knowledge management* (2nd ed.; pp. 354–365). Hershey, PA: IGI Global. doi:10.4018/978-1-59904-931-1.ch034

Zyngier, S. (2011). Knowledge management: Realizing value through governance. *International Journal of Knowledge Management, 7*(1), 35–54. doi:10.4018/jkm.2011010103

Zyngier, S. (2013). Knowledge management: Realizing value through governance. In M. Jennex (Ed.), *Dynamic models for knowledge-driven organizations* (pp. 36–55). Hershey, PA: IGI Global. doi:10.4018/978-1-4666-2485-6.ch003

Compilation of References

A Survey on Feature Selection Techniques and Classification Algorithms for Efficient Text Classification. (2016). *International Journal of Science and Research.*

Abdalla, M., Bellare, M., Catalano, D., Kiltz, E., Kohno, T., Lange, T., ... Shi, H. (2008). Searchable encryption revisited: Consistency properties, relation to anonymous ibe, and extensions. *Journal of Cryptology, 21*(3), 350–391. doi:10.100700145-007-9006-6

Alavi, A. H., & Gandomi, A. H. (2011). A robust data mining approach for formulation of geotechnical engineering systems. *Engineering Computations, 28*(3), 242–274. doi:10.1108/02644401111118132

Ali AlShaari, M. (2014). Text Documents Classification Using Word Intersections. *International Journal of Engineering and Technology, 6*(2), 119–122. doi:10.7763/IJET.2014.V6.678

Alicherry, M., & Lakshman, T. V. (2012). Network aware resource allocation in distributed clouds. *Proceedings - IEEE INFOCOM, 12*, 963–971.

Alliance, C. S. (2009). *Security Guidance for Critical Areas of Focus in Cloud Computing*. Retrieved from http://www.cloudsecurityalliance.org

Armbrust, Fox, Griffith, Joseph, & Katz, Konwinski, … Zaharia. (2010). A view of cloud computing. *Communications of the ACM, 53*(4), 50–58.

Arvor, D., Jonathan, M., Meirelles, M. S. P., Dubreuil, V., & Durieux, L. (2011). Classification of MODIS EVI time series for crop mapping in the state of MatoGrosso. *Brazil International Journal of Remote Sensing, 32*(22), 7847–7871. doi:10.1080/01431161.2010.531783

Austin. (2012). *eDiscovery Trends: CGOCs Information Lifecycle Governance Leader Reference Guide*. Retrieved from http://www.ediscoverydaily.com

Bakshi, K. (2012). Considerations for big data: Architecture and approach. *Aerospace Conference*, 1–7. 10.1109/AERO.2012.6187357

Ballard, L., Kamara, S., & Monrose, F. (2005). Achieving efficient conjunctive keyword searches over encrypted data. *Proc. of ICICS*. 10.1007/11602897_35

Barham, P., Donnelly, A., Isaacs, R., & Mortier, R. (2004). Using magpie for request extraction and workload modelling. *Proceedings of the 6th conference on Symposium on Opearting Systems Design & Implementation (OSDI'04)*, 18–18.

Basori, Afif, Almazyad, Abujabal, Rehman, & Alkawaz. (2015). Fast Markerless Tracking for Augmented Reality in Planar Environment. *3D Research, 6*(4), 1-11.

Bellare, M., Boldyreva, A., & Neill, A. O. (2007). Deterministic and efficiently searchable encryption. *Proc. of CRYPTO*.

Berry, M. J., & Linoff, G. S. (1997). *Data mining techniques: For marketing, sales, and customer support*. John Wiley & Sons, Inc.

Berzal, F., Cubero, J. C., Cuenca, F., & Martín-Bautista, M. J. (2003). On the quest for easy-to-understand splitting rules. *Data & Knowledge Engineering, 44*(1), 31–48. doi:10.1016/S0169-023X(02)00062-9

Beyer & Laney. (2012). *The importance of 'big data': A definition*. Gartner.

Big Data, Big Impact: New Possibilities for International Development. (n.d.). World Economic Forum.

Bijalwan, V., Kumar, V., Kumari, P., & Pascual, J. (2014). KNN based machine learning approach for text and document mining. *International Journal of Database Theory and Application, 7*(1), 61–70. doi:10.14257/ijdta.2014.7.1.06

Bill Franks. (2012). *Taming the big data tidal wave*. Wiley.

Boldyreva, A., Chenette, N., Lee, Y., & Oneill, A. (2009). Order-preserving symmetric encryption. In *Advances in Cryptology-EUROCRYPT* (pp. 224–241). Springer.

Boneh, Kushilevitz, Ostrovsky, & W. E. S. III. (2007). Public key encryption that allows pir queries. *Proc. of CRYPTO*.

Boneh, D., Crescenzo, G. D., Ostrovsky, R., & Persiano, G. (2004). Public key encryption with keyword search. *Proc. of EUROCRYPT*.

Boneh, D., & Waters, B. (2007). Conjunctive, subset, and range queries on encrypted data. *Proc. of TCC*, 535–554.

Bowyer, K. W., Hall, L. O., Moore, T., Chawla, N., & Kegelmeyer, W. P. (2000), A parallel decision tree builder for mining very large visualization datasets. *IEEE International Conference on Systems Man, and Cybernetics*, 3, 1888-1893. 10.1109/ICSMC.2000.886388

Brinkman, R. (2007). *Searching in encrypted data*. PhD thesis.

Brinkman. (2007). *Searching in encrypted data*. PhD thesis.

Cantrill, B. M., Shapiro, M. W., & Leventhal, A. H. (2004). Dynamic instrumentation of production systems. *USENIX Annual Technical Conference*, 15–28.

Cao, N., Wang, C., Li, M., Ren, K., & Lou, W. (2014). Privacy-preserving multikeyword ranked search over encrypted cloud data. *IEEE Transactions on Parallel and Distributed Systems*, *25*(1), 222–233. doi:10.1109/TPDS.2013.45

Cao, Wang, & Li, Ren, & Lou. (2014). Privacy-preserving multikeyword ranked search over encrypted cloud data. *IEEE Transactions on Parallel and Distributed Systems*, *25*(1), 222–233.

Cash, J., & Jutla, K., Ro3u, & Steiner. (2013). Highly-scalable searchable symmetric encryption with support for Boolean queries. *Proc. CRYPTO*, 353-373.

Chang, Y.-C., & Mitzenmacher, M. (2005). Privacy preserving keyword searches on remote encrypted data. *Proc. of ACNS*. 10.1007/11496137_30

Chen, M. Y., Accardi, A., Kiciman, E., Lloyd, J., Patterson, D., Fox, A., & Brewer, E. (2004). Path-based faliure and evolution management. In *Proceedings of the 1st conference on Symposium on Networked Systems Design and Implementation*. USENIX Association.

Chen, M., Kiciman, E., Fratkin, E., Fox, A., & Brewer, E. (n.d.). Pinpoint: problem determination in large, dynamic internet services. *Proceedings of the International Conference on Dependable Systems and Networks (DSN'02)*, 595–604. 10.1109/DSN.2002.1029005

Chen, X., Zheng, Z., Liu, X., Huang, Z., & Sun, H. (2011). *Personalized QoS-aware Web service recommendation and visualization. IEEE Transactions on Services Computing*.

Cloudera's 100% Open Source Distribution of Hadoop. (n.d.). Retrieved from http://www.cloudera.com/content/clou dera/en/products/cdh.html

Coyle, D. J., Jr., Chang, A., Malkemus, T. R., & Wilson, W. G. (1997). *U.S. Patent No. 5,630,124*. Washington, DC: U.S. Patent and Trademark Office.

Curtmola, R., Garay, J. A., Kamara, S., & Ostrovsky, R. (2006). Searchable symmetric encryption: improved definitions and efficient constructions. *Proc. of ACM CCS*. 10.1145/1180405.1180417

David, R. (2013). *Getting started with business analytics – insightful decision making*. Talor & Francis Group.

Dayan, P. (n.d.). Unsupervised Learning. Appeared. In R. A. Wilson & F. Keil (Eds.), *The MIT Encyclopedia Of The Cognitive Sciences*. MIT.

De Capitani di Vimercati, S., Foresti, S., & Samarati, P. (2012). Managing and accessing data in the cloud: Privacy risks and approaches. CRiSIS, 1 –9.

Dean, J., & Ghemawat, S. (2008). MapReduce: Simplified data processing on large clusters. *Communications of the ACM*, *51*(1), 107–113. doi:10.1145/1327452.1327492

DeCandia, Hastorun, Jampani, Kakulapati, Lakshman, Pilchin, … Vogels. (2007). Dynamo: amazon's highly available key-value store. *Proc. 21st ACM Symposium on Operating Systems Principles (SOSP'07)*, 205–220.

Demirkan, H., & Delen, D. (2013). Leveraging the capabilities of service-oriented decision support systems: Putting analytics and big data in cloud. *Decision Support Systems*, *55*(1), 412–421. doi:10.1016/j.dss.2012.05.048

Dey, A. (2016). A Systematic Review of Usability Studies in Augmented Reality between 2005 and 2014. *Mixed and Augmented Reality (ISMAR-Adjunct),* 2016 *IEEE International Symposium on*.

Dinh, Lee, Niyato, & Wang. (2013). A survey of mobile cloud computing: Architecture, applications, and approaches. *Wireless Commun. Mobile Comput.*, *13*(18).

Dong, X. L., & Srivastava, D. (2013). Big data integration. In *Data Engineering (ICDE), 2013 IEEE 29th International Conference on*, 1245–1248. 10.1109/ICDE.2013.6544914

Dong, J., Xiao, X., Kou, W., Qin, Y., Zhang, G., Li, L., ... Moore, B. III. (2015). Tracking the dynamics of paddy rice planting area in 1986-2010 through time series Landsat images and phenology-based algorithms. *Remote Sensing of Environment*, *160*, 99–113. doi:10.1016/j.rse.2015.01.004

Dubey, V. (2016). *Hybrid classification model of correlation-based feature selection and support vector machine*. IEEE. doi:10.1109/ICCTAC.2016.7567338

Elnikety, S., Pedone, F., & Zwaenepoel, W. (2005, October). Database replication using generalized snapshot isolation. In *Reliable Distributed Systems, 2005. SRDS 2005. 24th IEEE Symposium on* (pp. 73-84). IEEE. 10.1109/RELDIS.2005.14

Elnikety, S., Dropsho, S., & Pedone, F. (2006, April). Tashkent: Uniting durability with transaction ordering for high-performance scalable database replication. *Operating Systems Review*, *40*(4), 117–130. doi:10.1145/1218063.1217947

Eswaran, K. P., Gray, J. N., Lorie, R. A., & Traiger, I. L. (1976). The notions of consistency and predicate locks in a database system. *Communications of the ACM*, *19*(11), 624–633. doi:10.1145/360363.360369

Gayathri, K. (2013). *Data pre-processing with the KNN for classification using the SVM*. IEEE.

Gentry, C. (2009). *A fully homomorphic encryption scheme* (Ph.D. dissertation). Stanford University.

Global data center traffic – Cisco Forecast Overview. (n.d.). Retrieved from http://www.cisco.com/en/US/solutions/ collateral/ns341/ns525/ns537/ns705/ns 1175/Cloud_Index_White_Paper.html

Goh, E.-J. (2003). *Secure indexes*. Retrieved from http://eprint.iacr.org/2003/216

Goldreich, O., & Ostrovsky, R. (1996). Software protection and simulation on oblivious rams. *Journal of the Association for Computing Machinery*, *43*(3), 431–473. doi:10.1145/233551.233553

Golle, P., Staddon, J., & Waters, B. (2004). Secure conjunctive keyword search over encrypted data. *Proc. of ACNS*, 31–45. 10.1007/978-3-540-24852-1_3

Graham, S. L., Kessler, P. B., & Mckusick, M. K. (1982). Gprof: A call graph execution profiler. *ACM SIGPLAN Notices*, *17*(6), 120–126. doi:10.1145/872726.806987

Guang-chao, W. (2008). Support Vector Machine Classifier Based on Fuzzy Partition and Neighborhood Pairs. *Jisuanji Yingyong*.

Gumus, F. (2014). *Online Naive Bayes classification for network intrusion detection*. IEEE. doi:10.1109/ASONAM.2014.6921657

Gunaratna, Kodikara, & Premaratne. (2008). ANN based currency recognition system using compressed gray scale and application for Sri Lankan currency notes-SLCRec. *Proceedings of World Academy of Science, Engineering and Technology, 35*, 235-240.

Guo, Zhao, & Cai. (2010). A reliable method for paper currency recognition based on LBP. In *Network Infrastructure and Digital Content, 2010 2nd IEEE International Conference on*. IEEE.

Gupta, A., Agarwal, D., Tan, D., Kulesza, J., Pathak, R., Stefani, S., & Srinivasan, V. (2015, May). Amazon Redshift and the case for simpler data warehouses. In *Proceedings of the 2015 ACM SIGMOD International Conference on Management of Data* (pp. 1917-1923). ACM. 10.1145/2723372.2742795

Han, S., Dang, Y., Ge, S., Zhang, D., & Xie, T. (2012). Performance debugging in the large via mining millions of stack traces. *Proc. 34th Int'l Conf. on Software Engineering (ICSE'12)*, 145–155. 10.1109/ICSE.2012.6227198

Hassanpour, Yaseri, & Ardeshiri. (2007). Feature extraction for paper currency recognition. In *Signal Processing and Its Applications,* 2007. *ISSPA 2007. 9th International Symposium on*. IEEE.

Hitzler, P., & Janowicz, K. (2013). Linked Data, Big Data, and the 4th Paradigm. *Semantic Web*, *4*(3), 233–235.

Howe, D., Costanzo, M., Fey, P., Gojobori, T., Hannick, L., Hide, W., ... Rhee, S. Y. (2008). Big data: The future of biocuration. *Nature*, *455*(7209), 47–50. doi:10.1038/455047a PMID:18769432

Hwan, Gelogo, & Park. (2012). Next Generation Cloud Computing Issues and Solutions. *International Journal of Control and Automation, 5*.

Hwang, Y., & Lee, P. (2007). *Public key encryption with conjunctive keyword search and its extension to a multi-user system*. Pairing. doi:10.1007/978-3-540-73489-5_2

IBM. (2013). *What is big data? ł bringing big data to the enterprise*. Retrieved from http://www-01.ibm.com/software/data/bigdata

Jason, J. K. P., Park, Y., & Mahlke, S. (2013). Efficient execution of augmented reality applications on mobile programmable accelerators. *Field-Programmable Technology (FPT) 2013 International Conference on*, 176-183.

Jung, T., Mao, X., Li, X., Tang, S.-J., Gong, W., & Zhang, L. (2013). Privacy preserving data aggregation without secure channel: multivariate polynomial evaluation. *Proceedings of INFOCOM*, 2634–2642. 10.1109/INFCOM.2013.6567071

Kamara & Lauter. (2010). Cryptographic cloud storage. In *RLCPS*. Springer.

Kamara, S., & Lauter, K. (2010). Cryptographic cloud storage. In RLCPS. Springer. doi:10.1007/978-3-642-14992-4_13

Kang, Y., Zheng, Z., & Lyu, M. (2012). A latency-aware co-deployment mechanism for cloud-based services. *Proceedings of the IEEE 5th International Conference on Cloud Computing (CLOUD'12)*, 630–637. 10.1109/CLOUD.2012.90

Katz, J., Sahai, A., & Waters, B. (2008). *Predicate encryption supporting disjunctions, polynomial equations, and inner products. Proc. of EUROCRYPT*. doi:10.1007/978-3-540-78967-3_9

Khan, Wang, Kulsoom, & Ullah. (2013). Searching Encrypted Data on Cloud. *International Journal of Computer Science Issues, 10*(6).

Kristensen, B. B., May, D., & Nowack, P. (2011). Collaboration and Modeling in Ambient Systems: Vision Concepts and Experiments. *System Sciences (HICSS) 2011 44th Hawaii International Conference on*, 1-6.

Kulesza, A. (2012). Determinantal Point Processes for Machine Learning. Foundations And Trends® In. *Machine Learning*, *5*(2-3), 123–286. doi:10.1561/2200000044

Kumar, S. (2004, January 01). MEGA3: Integrated software for Molecular Evolutionary Genetics Analysis and sequence alignment. *Briefings in Bioinformatics*, *5*(2), 150–163. doi:10.1093/bib/5.2.150 PMID:15260895

Kumar, V., Grama, A., Gupta, A., & Karypis, G. (1994). *Introduction to parallel computing* (Vol. 110). Redwood City: Benjamin/Cummings.

Kumar, V., Grama, A., Gupta, A., & Karypis, G. (1994). *Introduction to parallel computing: design and analysis of algorithms* (Vol. 400). Redwood City, CA: Benjamin/Cummings.

Kuttiyapillai, D., & Ramachandran, R. 2014. Design and analysis of feature classification model using information extraction in tomato growing environment. International Information Institute (Tokyo). Information, 17(8), 3947-3959.

Kuttiyapillai, D., & Rajeswari, R. (2015). A method for extracting task-oriented information from biological text sources. *International Journal of Data Mining and Bioinformatics*, *12*(4), 387–399. doi:10.1504/IJDMB.2015.070072 PMID:26510293

Labrinidis, A., & Jagadish, H. V. (2012). Challenges and opportunities with big data. *Proceedings of the VLDB Endowment International Conference on Very Large Data Bases*, *5*(12), 2032–2033. doi:10.14778/2367502.2367572

Laskov, P., & Lippmann, R. (2010). Machine learning in adversarial environments. *Machine Learning*, *81*(2), 115–119. doi:10.100710994-010-5207-6

Lewko, A., Okamoto, T., Sahai, A., Takashima, K., & Waters, B. (2010). Fully secure functional encryption: Attribute-based encryption and (hierarchical) inner product encryption. *Proc. of EUROCRYPT*. 10.1007/978-3-642-13190-5_4

Li, H., Dai, Y., Tian, L., & Yang, H. (2009). Identity-based authentication for cloud computing. In Cloud Computing. Berlin, Germany: Springer-Verlag. doi:10.1007/978-3-642-10665-1_14

Li, J., Wang, Q., Wang, C., Cao, N., Ren, K., & Lou, W. (2010). Fuzzy keyword search over encrypted data in cloud computing. Proc. of IEEE INFOCOM'10 Mini-Conference. doi:10.1109/INFCOM.2010.5462196

Liang, Cai, Huang, Shen, & Peng. (2012). An SMDP-based service model for interdomain resource allocation in mobile cloud networks. *IEEE Trans. Veh. Technol.*, *61*(5).

Liang, H., Cai, L. X., Huang, D., Shen, X., & Peng, D. (2012). An smdpbased service model for interdomain resource allocation in mobile cloud networks. *IEEE Transactions on Vehicular Technology*, *61*(5), 2222–2232. doi:10.1109/TVT.2012.2194748

Li, H., Liu, D., Dai, Y., Luan, T. H., & Shen, X. (2014). Enabling efficient multi-keyword ranked search over encrypted cloud data through blind storage. *IEEE Transactions on Emerging Topics in Computing*. doi:10.1109/TETC.2014.2371239

Lim, F., Singh, N., & Yajnik, S. (2008). A log mining approach to failure analysis of enterprise telephony systems. *Proceedings of the IEEE International Conference on Dependable Systems and Networks (DSN'08)*, 398–403.

Li, R., Xu, Z., Kang, W., Yow, K. C., & Xu, C.-Z. (2014). Efficient multikeyword ranked query over encrypted data in cloud computing. *Future Generation Computer Systems*, *30*, 179–190. doi:10.1016/j.future.2013.06.029

Liu, Li, Huang, & Wen. (2012). Shapley value based impression propagation for reputation management in web service composition. *Pro. IEEE 19th Int'l Conf' on Web Services (ICWS'12)*, 58–65.

Lo, W., Yin, J., Deng, S., Li, Y., & Wu, Z. (2012). Collaborative web service qos prediction with location-based regularization. *Pro. IEEE 19th Int'l Conf' on Web Services (ICWS'12)*, 464–471. 10.1109/ICWS.2012.49

Lobell, D. B., Thau, D., Seifert, C., Engle, E., & Little, B. (2015). A scalable satellite-based crop yield mapper. *Remote Sensing of Environment*, *164*, 324–333. doi:10.1016/j.rse.2015.04.021

Lyu, M. R. (1995). *Software Fault Tolerance. In Trends in Software*. Wiley.

MacWilliams, Reicher, Klinker, & Bruegge. (2014). Design Patterns for Augmented Reality Systems. *CEUR Workshop Proceedings*, 19.

Madsen, J. B., & Stenholt, R. (2014). How wrong can you be: Perception of static orientation errors in mixed reality. *3D User Interfaces (3DUI) 2014 IEEE Symposium on*, 83-90. 10.1109/3DUI.2014.6798847

Mahmoud & Shen. (2012). A cloud-based scheme for protecting source-location privacy against hotspot-locating attack in wireless sensor networks. *IEEE Trans. Parallel Distrib. Syst.*, *23*(10).

Mahmoud, M. M., & Shen, X. (2012). A cloud-based scheme for protecting source-location privacy against hotspot-locating attack in wireless sensor networks. *IEEE Transactions on Parallel and Distributed Systems*, *23*(10), 1805–1818. doi:10.1109/TPDS.2011.302

Manyika, Chui, Brown, Bughin, Dobbs, Roxburgh, & Byers. (2011). *Big data: The next frontier for innovation, competition, and productivity*. McKinsey Global Institute.

Mark, A. (2012). *Beyer and Douglas Laney. "The Importance of 'Big Data': A Definition*. Gartner.

Mell & Grance. (2011). The nist definition of cloud computing (draft). *NIST Special Publication, 800*, 145.

Metropolis, N. (1986). Massively parallel processing. *Journal of Scientific Computing*, *1*(2), 115–116. doi:10.1007/BF01061388

Mi, H., Wang, H., Cai, H., Zhou, Y., Lyu, M. R., & Chen, Z. (2012). Ptracer: Path-based performance profiling in cloud computing systems. In *Proceedings of the 36th IEEE Annual Computer Software and Applications Conference (COMPSAC'12)*. IEEE.

Minka, T. P., & Picard, R. W. (1997). Interactive learning using a "society of models". *Pattern Recognition*, *30*(4), 565–581. doi:10.1016/S0031-3203(96)00113-6

Mi, Wang, & Zhou, Lyu, & Cai. (2013). Towards fine-grained, unsupervised, scalable performance diagnosis for production cloud computing systems. *IEEE Transactions on Parallel and Distributed Systems*.

Naveed, Prabhakaran, & Gunter. (2014). Dynamic searchable encryption via blind storage. *Proceedings - IEEE Symposium on Security and Privacy*, 639–654.

Nguyen, B. V., Pham, D., Ngo, T. D., Le, D. D., & Duong, D. A. (2014, December). Integrating spatial information into inverted index for large-scale image retrieval. In *Multimedia (ISM), 2014 IEEE International Symposium on* (pp. 102-105). IEEE. doi:10.1007/978-3-319-12024-9_19

OECD. (2015). *Data-driven innovation: big data for growth and well-being*. Paris, France: OECD Publishing.

Oracle Big Data strategy guide. (n.d.). Retrieved from http://www.oracle.com/us/technologies /big-data/big-data-strategy-guide- 1536569.pdf

Paillier, P. (1999). Public key cryptosystems based on composite degree residuosity classes. Eurocrypt, 223–238. doi:10.1007/3-540-48910-X_16

Pang, Shen, & Krishnan. (n.d.). Privacy-preserving similarity-based text retrieval. ACM Transactions on Internet Technology, 10(1), 4.

Polo, J. L., Berzal, F., & Cubero, J. C. (2007). Taking class importance into account. *Lecture Notes in Computer Science*, 4413.

Provost & Fawcett. (2013). *Data science for business*. O'Reilly.

Quinlan. (1996). *Improved use of continuous attributes in C4.5*. arXiv preprint cs/9603103

Quinlan, J. R. (1986). Induction of decision trees. *Machine Learning*, *1*(1), 81–106. doi:10.1007/BF00116251

Quinlan, J. R. (1990). Decision trees and decision-making. *IEEE Transactions on Systems, Man, and Cybernetics*, *20*(2), 339–346. doi:10.1109/21.52545

Quinlan, J. R. (1993). *C4.5: programs for machine learning*. Morgan Kaufmann.

Ramesh, B. (2015). Big data architecture. In *Big Data* (pp. 29–59). Springer. doi:10.1007/978-81-322-2494-5_2

Reddy. (2013). Techniques for Efficient Keyword Search in Cloud Computing. *International Journal of Computer Science and Information Technologies, 4*(1).

Reges, S., & Stepp, M. (2014). *Building Java Programs*. Pearson.

Ren, K., Wang, C., & Wang, Q. (2012). Security Challenges for the Public Cloud. *IEEE Internet Computing*, *16*(1), 69–73. doi:10.1109/MIC.2012.14

Reynolds, P., Killian, C., Wiener, J. L., Mogul, J. C., Shah, M. A., & Vahdat, A. (2006). Pip: detecting the unexpected in distributed systems. *Proceedings of the 3rd conference on Networked Systems Design & Implementation (NSDI'06)*, 9–9.

Ristoski, P., Mencía, E. L., & Paulheim, H. (2014, May). A hybrid multi-strategy recommender system using linked open data. In Semantic Web Evaluation Challenge (pp. 150-156). Springer International Publishing. doi:10.1007/978-3-319-12024-9_19

Ristoski, P., Mencía, E. L., & Paulheim, H. (2014, May). A hybrid multi-strategy recommender system using linked open data. In *Semantic Web Evaluation Challenge* (pp. 150–156). Springer International Publishing.

Rui, Y., Huang, T. S., Ortega, M., & Mehrota, S. (1998). Relevance feedback: A power tool for interactive content-based image retrieval. *IEEE Transaction on Circuits System and Video Technnology*, *8*(5), 644–655. doi:10.1109/76.718510

Samanthula, B. K., Elmehdwi, Y., & Jiang, W. (2014). *k-nearest neighbor classification over semantically secure encrypted relational data*. eprint arXiv:1403.5001

Sankaranarayanan, S., & Perumal, T. P. (2014). Diabetic Prognosis through Data Mining Methods and Techniques," *Intelligent Computing Applications (ICICA), 2014 International Conference on*, 162-166.

Schatz, M. C. (2009). CloudBurst: Highly sensitive read mapping with MapReduce. *Bioinformatics (Oxford, England)*, *25*(11), 1363–1369. doi:10.1093/bioinformatics/btp236 PMID:19357099

Senthil, G. (2004). *U.S. Patent No. 6,721,869*. Washington, DC: U.S. Patent and Trademark Office.

Shafer, J., Agrawal, R., & Mehta, M. (1996). PRINT: A scalable parallel classifier for data mining. *Proc. Int.Conf. Very Large Data Bases*.

Shao, L., Zhang, J., Wei, Y., Zhao, J., Xie, B., & Mei, H. (2007). Personalized QoS prediction for Web services via collaborative filtering. *Proc. 5th Int'l Conf. Web Services (ICWS'07)*, 439–446. 10.1109/ICWS.2007.140

Shen, Liang, Shen, Lin, & Luo. (2014). Exploiting geodistributed clouds for a e-health monitoring system with minimum service delay and privacy preservation. *IEEE J. Biomed. Health Inform.*, *18*(2).

Shen, E., Shi, E., & Waters, B. (2009). Predicate privacy in encryption systems. *Proc. of TCC*.

Shen, Q., Liang, X., Shen, X., Lin, X., & Luo, H. (2014). Exploiting geodistributed clouds for e-health monitoring system with minimum service delay and privacy preservation. *IEEE Journal of Biomedical and Health Informatics*, *18*(2), 430–439. doi:10.1109/JBHI.2013.2292829 PMID:24608048

Sigelman, Barroso, Burrows, Stephenson, Plakal, Beaver, … Shanbhag. (2010). *Dapper, a large-scale distributed systems tracing infrastructure*. Google, Inc.

Singhal, A. (2001). Modern information retrieval: A brief overview. *A Quarterly Bulletin of the Computer Society of the IEEE Technical Committee on Data Engineering*, *24*(4), 35–43.

Song, D. X., Wagner, D., & Perrig, A. (2000). Practical techniques for searches on encrypted data. In *Proceedings of S&P.* IEEE.

Song, D. X., Wagner, D., & Perrig, A. (2000). Practical techniques for searches on encrypted data. *Proceedings of S&P*, 44–55.

Song, D., Wagner, D., & Perrig, A. (2000). Practical techniques for searches on encrypted data. *Proc. of S&P*.

Stefanov, Papamanthou, & Shi. (2014). Practical dynamic searchable encryption with small leakage. *Proc. NDSS*. doi:10.1109/TPDS.2013.282

Steiner, M., Gaglianello, B. G., Gurbani, V. K., Hilt, V., Roome, W. D., Scharf, M., & Voith, T. (2012). Network-aware service placement in a distributed cloud environment. *Proc. ACM SIGCOMM'*12, 73–74. 10.1145/2342356.2342366

Sun, W. (2013). Privacy-preserving multi-keyword text search in the cloud supporting similarity-based ranking. Proc. 8th ACM SIGSAC Symp.Inf., Comput. Commun. Secur., 71-82.

Sun, Wang, Cao, Li, Lou, Hou, & Li. (2013). Verifiable privacy-preserving multi-keyword text search in the cloud supporting similarity-based ranking. *IEEE Transactions on Parallel and Distributed Systems*. DOI: 10.1109/TPDS.2013.282

Talwar, A., & Kumar, Y. (2013). Article. *International Journal of Engineering and Computer Science, 2*(12).

Tang, M., Jiang, Y., Liu, J., & Liu, X. F. (2012). Location-aware collaborative filtering for qos-based service recommendation. *Pro. IEEE 19th Int'l Conf' on Web Services (ICWS'12)*, 202–209.

Taylor, R. C. (2010). An overview of the Hadoop/MapReduce/HBase framework and its current applications in bioinformatics. *BMC Bioinformatics*, 11. PMID:21210976

Teather & Stuerzlinger. (2016). SIVARG: Spatial Interaction in Virtual/Augmented Reality and Games. Proceedings of the 2016. doi:10.1109/ISMAR-Adjunct.2016.0036

The Economist. (2010). A special report on managing information: Data, data everywhere. *The Economist*.

Thereska, G., Salmon, B., Strunk, J., Wachs, M., Abd-El-Malek, M., Lopez, J., & Ganger, G. R. (2006). Stardust: tracking activity in a distributed storage system. ACM SIGMETRICS Performance Evaluation Review, 34(1), 3–14. doi:10.1145/1140277.1140280

Thereska, H., & Ganger, G. R. (2008). Ironmodel: robust performance models in the wild. *Proceedings of the International Conference on Measurement and Modeling of Computer Systems (SIGMETRICS '08)*, 253–264. 10.1145/1375457.1375486

Tukey, J. W. (1977). *Exploratory data analysis*. Academic Press.

Valsala, S., Ann George, J., & Parvathy, P. (2011). A Study of Clustering and Classification Algorithms Used in Datamining. *International Journal of Computer Science and Network Security, 11*(10).

Vaquero, L. M., Rodero-Merino, L., Caceres, J., & Lindner, M. (2009). A break in the clouds: Towards a cloud definition. *ACM SIGCOMM Comput. Commun. Rev., 39*(1), 50–55. doi:10.1145/1496091.1496100

Veltkamp, C. R., Burkhardt, H., & Kriegel, H. P. (2001). *State-of-the-Art in content-based image and video retrieval*. Norwell, MA: Kluwer. doi:10.1007/978-94-015-9664-0

Venkatesan, Arunkumar, Thangavelu, & Prabhavathy. (2013). An Improved Bayesian Classification Data mining Method for Early Warning Landslide Susceptibility Model Using GIS. *Proceedings of Seventh International Conference on Bio-Inspired Computing: Theories and Applications (BIC-TA 2012)*. Springer.

Venkatesan, Rajawat, Arunkumar, Anbarasi, & Malarvizhi. (2014). GIS Based Data Mining Classification Approaches for Landslide Susceptibility Analysis. *International Journal of Applied Environmental Sciences*, *9*(5), 2345–2357.

Vermote, E. F., Tanre, D., Deuze, J. L., Herman, M., & Morcrette, J. (1997). Second simulation of the satellite signal in the solar spectrum, 6S: An overview. *IEEE Transactions on Geoscience and Remote Sensing*, *35*(3), 675–686. doi:10.1109/36.581987

Wang, C., Cao, N., Li, J., Ren, K., & Lou, W. (2010). Secure ranked keyword search over encrypted cloud data. In *Proceedings of ICDCS*. IEEE. 10.1109/ICDCS.2010.34

Wang, L. (2012). *Improved KNN classification algorithms research in text categorization*. IEEE. doi:10.1109/CECNet.2012.6201850

Wang, S. (2013). *Support Vector Machines Classification for High-Dimensional Dataset*. IEEE.

Wang, Yu, Lou, & Hou. (2014). Privacy-preserving multi-keyword fuzzy search over encrypted data in the cloud. *Proceedings - IEEE INFOCOM*.

White. (2012). *Hadoop the definitive guide*. O'Reilly Media, Inc.

Williams, P., Sion, R., & Carbunar, B. (2008). Building castles out of mud: practical access pattern privacy and correctness on untrusted storage. ACM CCS, 139–148. doi:10.1145/1455770.1455790

Witten, H. I., & Frank, E. (2005). Data Mining: Practical machine learning tools and techniques. Morgan Kaufmann.

Witten, Moffat, & Bell. (1999). *Managing gigabytes: Compressing and indexing documents and images*. Morgan Kaufmann Publishing.

Witten, Moffat, & Bell. (1999). *Managing Gigabytes: Compressing and indexing documents and images*. Morgan Kaufmann Publishing.

Wong, W. K., Cheung, D. W., Kao, B., & Mamoulis, N. (2009). Secure knn computation on encrypted databases. *Proc. of SIGMOD*. 10.1145/1559845.1559862

Wong, W. K., Cheung, D. W., Kao, B., & Mamoulis, N. (2010). Secure kNN computation on encrypted databases. Proc. ACM SIGMOD Int. Conf. Manage. Data, 139-152.

Wu, G. (2008). Support vector machine classifier based on fuzzy partition and neighborhood pairs. *Jisuanji Yingyong*, *28*(1), 131–133. doi:10.3724/SP.J.1087.2008.00131

Wu, H.-K., Lee, S. W.-Y., Chang, H.-Y., & Liang, J.-C. (2013, March). Current status, opportunities and challenges of augmented reality in education. *Computers & Education*, *62*(C), 41–49. doi:10.1016/j.compedu.2012.10.024

Wu, Zhu, & Wu, & Ding. (2014). Data Mining with Big Data. *IEEE Transactions on Knowledge and Data Engineering*, *26*(1).

Xu, W., Huang, L., Fox, A., Patterson, D., & Jordan, M. I. (2009). Detecting large-scale system problems by mining console logs. *Proceedings of the ACM 22nd Simposium on Operating Systems Principles (SOSP'09)*, 117–132. 10.1145/1629575.1629587

Yang, L., Liu, & Yang. (2014). Secure dynamic searchable symmetric encryption with constant document update cost. *Proc.GLOBECOM*.

Yang, Li, Liu, Yang, & M. (2014). Secure dynamic searchable symmetric encryption with constant document update cost. In *Proceedings of GLOBCOM*. IEEE.

Yang, H., Xu, Z., King, I., & Lyu, M. (2010). Online learning for group lasso. *International Conference on Machine Learning (ICML'10)*.

Yi, X., Liu, F., Liu, J., & Jin, H. (2014). Building a network highway for big data: Architecture and challenges. *IEEE Network*, *28*(4), 5–13. doi:10.1109/MNET.2014.6863125

Yu, J., Lu, P., Zhu, Y., Xue, G., & Li, M. (2013). Towards secure multikeyword top-k retrieval over encrypted cloud data. *IEEE Transactions on Dependable and Secure Computing*, *10*(4), 239–250. doi:10.1109/TDSC.2013.9

Zhang, E.-H. (2003). Research on paper currency recognition by neural networks. In *Machine Learning and Cybernetics, 2003 International Conference on* (vol. 4). IEEE.

Zhang, B., & Zhang, F. (2011). An efficient public key encryption with conjunctive-subset keywords search. *Journal of Network and Computer Applications*, *34*(1), 262–267. doi:10.1016/j.jnca.2010.07.007

Zhang, Q., Zhu, Q., Zhani, M. F., & Boutaba, R. (2012). Dynamic service placement in geographically distributed clouds. *Proc. IEEE 32nd Int'l Conf. on Distributed Computing Systems (ICDCS'12)*, 526–535. 10.1109/ICDCS.2012.74

Zheng, Z., & Lyu, M. R. (2009). A QoS-aware fault tolerant middleware for dependable service composition. *Proc. 39th Int'l Conf. Dependable Systems and Networks (DSN'09)*, 239–248. 10.1109/DSN.2009.5270332

Zheng, Z., Ma, H., Lyu, M. R., & King, I. (2011). QoS-aware Web service recommendation by collaborative filtering. *IEEE Transactions on Services Computing*, *4*(2), 140–152. doi:10.1109/TSC.2010.52

Zheng, Z., Zhang, Y., & Lyu, M. R. (2010). CloudRank: A QoS-driven component ranking framework for cloud computing. *Proc. Int'l Symp. Reliable Distributed Systems (SRDS'10)*, 184–193. 10.1109/SRDS.2010.29

Zheng, Z., Zhu, J., & Lyu, M. R. (2013). Usage-created Big Data and Big Data-as-a-Service: An Overview. *IEEE International Congress on Big Data.*

Zhu, J., Zheng, Z., Zhou, Y., & Lyu, M. R. (2013). Scaling service-oriented applications into geo-distributed clouds. *Pro. IEEE Int'l Workshop on Internet-based Virtual Computing Environment (iVCE'13).*

Zikopoulos, P., & Eaton, C. (2011). *Understanding big data: Analytics for enterprise class hadoop and streaming data*. McGraw-Hill Osborne Media.

About the Contributors

Leonid Datta is a final year student of VIT University (Computer Science and Engineering). Completed projects entitled "truncation technique for search engine", "Web design and data analysis" and "Universal Student Identification using RFID Sensor".

Sameera K. is a Research Associate, School of Computer Science and Engineering, VIT University.

Chetan Kumar is a student of VIT University.

Venkatesan M. is an Assistant Professor in Department of Computing Science and Engineering, National Institute of Technology Karnataka, Mangalore. He is working in the area of big data analytics, data science and spatial data mining. His research area includes databases, data warehouse, data mining, big data and applications of data mining in various real-time applications. He has good knowledge in data mining tools like Clementine and Rapidminer. He is the Principal Investigator in ISRO funded project on data mining in a landslide. He holds a BE, MTech and PhD in the area of spatial data mining.

Abhishek Mukherjee is a student of VIT University.

Prabhavathy P. is working as Associate Professor in School of Information Technology and Ennginneering, VIT University, Vellore. She has completed PhD in the area of computational Intelligence. Her area of interest includes rough set, fuzzy set, Sequential mining, data mining. She published number of paper in international journals and international conference.

Pronay Peddiraju is a Computer Science Engineer with experience in game engine technology, Augmented and Virtual Reality development and 3D object visualisation. Pronay has performed research in fields of game technology and problem solving.

Dhanasekaran Pillai received his Ph. D in Computer Science and Engineering under the Faculty of Information and Communication Engineering from the Anna University in 2015, India. In 2009 he was received M.E. in Computer Science and Engineering from the Anna University and is currently the Associate Professor at JGI-Jain College of Engineering, Belgaum, Karnataka, India. He writes and presents widely on issues of data mining, information extraction, and big data analytics, remote sensing, and is the member of Editorial and reviewer board of IJRCIS.

Remya S. is now doing research in VIT University, Vellore. Completed M.Tech in Calicut University during 2009-2011 and B.Tech in 2002-2006 from CUSAT.

Shweta Shinde is a VIT University student working in Genesys.

Ramasubbareddy Somula is now working as Assistant Professor in VIT University. Done BTech in Alfa College of Engineering & Technology, Andhra Pradesh. His MTech was in Srirama Engg College, AndhraPradesh.

Utkarsh Srivastava is a final year undergraduate from Computer Science background studying in VIT University. He is extremely passionate about learning and implementing technology. He also has deep interest in the field on Bigdata. A polyglot at heart, he can code and communicate in several languages.

Index

A

B

C

D

E

F

H

I

K

L

www.ingramcontent.com/pod-product-compliance
Lightning Source LLC
LaVergne TN
LVHW081325300926
849161LV00008B/1881